CLOSING
THE GAP

*Digital Equity Strategies
for Teacher Prep Programs*

NICOL R. HOWARD, SARAH THOMAS, REGINA SCHAFFER

International Society for Technology in Education
PORTLAND, OR • ARLINGTON, VA

Closing the Gap
Digital Equity Strategies for Teacher Prep Programs
Nicol R. Howard, Sarah Thomas, Regina Schaffer

© 2018 International Society for Technology in Education
World rights reserved. No part of this book may be reproduced or transmitted in any form or by any means—electronic, mechanical, photocopying, recording, or by any information storage or retrieval system—without prior written permission from the publisher. Contact Permissions Editor: iste.org/about/permissions-and-reprints; permissions@iste.org; fax: 1.541.302.3780.

Acquisitions Editor: *Valerie Witte*
Developmental and Copy Editor: *Linda Laflamme*
Proofreader: *Steffi Drewes*
Indexer: *Wendy Allex*
Book Design and Production: *Kim McGovern*
Cover Design: *Eddie Ouellette*

Library of Congress Cataloging-in-Publication Data available

First Edition
ISBN: 9781564847133

Ebook version available

Printed in the United States of America

ISTE® is a registered trademark of the International Society for Technology in Education.

About ISTE

The International Society for Technology in Education (ISTE) is the premier nonprofit organization serving educators and education leaders committed to empowering connected learners in a connected world. ISTE serves more than 100,000 education stakeholders throughout the world.

ISTE's innovative offerings include the ISTE Conference & Expo, one of the biggest, most comprehensive edtech events in the world—as well as the widely adopted ISTE Standards for learning, teaching and leading in the digital age and a robust suite of professional learning resources, including webinars, online courses, consulting services for schools and districts, books, and peer-reviewed journals and publications. Visit iste.org to learn more.

Join Our Community of Passionate Educators

ISTE members get free year-round professional development opportunities and discounts on ISTE resources and conference registration. Membership also connects you to a network of educators who can instantly help with advice and best practices.

Join or renew your ISTE membership today!

Visit iste.org/membership or call 800.336.5191.

Related ISTE Titles

Learning Supercharged: Digital Age Strategies and Insights from the Edtech Frontier, by Lynne Schrum with Sandi Sumerfield

To see all books available from ISTE, please visit iste.org/resources.

About the Authors

Nicol R. Howard, PhD is an assistant professor in the School of Education at the University of Redlands. She has served as co-chair for ISTE's Digital Equity Network, chair of the American Educational Research Association's Technology, Instruction, Cognition, and Learning SIG, and co-chair for the California Council on Teacher Education (CCTE) Technology SIG. Her research foci are teacher education, STEM and computer science opportunities for students of color, and the equitable uses of technology in K–16 classrooms. Her writing has appeared in the Corwin Connected Educators Series, *Urban Education Journal*, *International Journal of Educational Technology*, *Technology, Knowledge and Learning*, EDUCAUSE, Edutopia, and eCampus News. She is also the co-founder and co-editor of the *Journal of Computer Science Integration*.

Dr. Howard holds a Bachelor's degree in Sociology from the University of California at Los Angeles (UCLA), an MA in Educational Technology from Azusa Pacific University, and a PhD in Cultural and Curricular Studies from Chapman University. She is also a Google Education Trainer, a Microsoft Innovative Educator Trainer, and Raspberry Pi certified.

Sarah-Jane Thomas, PhD is a Regional Technology Coordinator in Prince George's County Public Schools. Sarah is also a Google Certified Innovator, Google Education Trainer, and founder of the EduMatch project, which promotes connection and collaboration

About the Authors

among educators around the world. Through EduMatch, Sarah has published several collaborative and individual books, and serves as president on the Board of Directors for EduMatch Foundation, Inc. Sarah is also on the leadership team of the ISTE Digital Equity PLN, and affiliate faculty at Loyola University in Maryland.

Sarah recently graduated from George Mason University with a doctoral degree in Education, and a concentration in International Education. She also holds a Master's degree from Howard University in the field of Curriculum and Instruction.

Sarah was designated as an ASCD Emerging Leader in 2016, and was named Prince George's County Public Schools Outstanding Educator Using Technology in 2015. She is also a winner of the 2014 Digital Innovation in Learning Award in the "Sharing is Caring" category, and was named by the National School Boards Association as one of the "20 to Watch" in 2015. She was part of the Technical Working Group that refreshed the International Society for Technology in Education (ISTE) Standards for Educators in 2016-2017, and in 2017 she received the ISTE Making IT Happen Award.

Regina Schaffer is a Technology Specialist for the Middletown Township School District. She was recently selected as a National School Boards Association's "20 to Watch," as well as a 2016 PBS LearningMedia Digital Innovator. She is a Google Certified Innovator, Common Sense Media Certified Teacher, Discovery Education DEN Star, Raspberry Pi certified, and a BrainPOP Connected Educator. Connect with Regina at @reginaschaffer.

Acknowledgments

We are thankful for the opportunity to author this book, and we are grateful for Valerie Witte's guidance throughout this journey as well as the support of ISTE's Digital Equity Leadership team from 2016 to present.

Special thanks to our contributors: Abigale Almerido, Patricia Brown, Dr. Marialice Curran, Rafranz Davis, Dr. Nirmla Flores, Matthew Heifield, Dr. Keith E. Howard, Dr. Betina Hsieh, Christine Hunt, Dr. Lara Kassab, and Dr. Stephanie Quan-Lorey.

Dedication

Continued thanks to my husband Keith and our children for their love and encouragement. Special thanks to my grandmother—our matriarch, Veronica Conant Metoyer—for modeling tremendous dedication to education for 25 years.

—Nicol

Thank you to my family. You are my everything. "Yay, yay, they are great." Much love always.

—Sarah

Thank you to my family, whose love and support continues to be the wind beneath my wings.

—Regina

Contents

INTRODUCTION 1
    What's in This Book 2
    Who This Book Is For 4

Chapter 1
CRITICAL DIGITAL EQUITY ISSUES 5
    Brief Overview of Digital Equity in K–12 Today 7
    Defining the Critical Issues in Teacher Preparation 15
    What Would You Do? 28
    DE Wisdom 28

Chapter 2
LEVERAGING MOBILE AND SOCIAL TRENDS 31
    The Horizon Report 32
    Mobile Devices 35
    Social Media 38
    Personal Broadcasting 43
    What Would You Do? 44
    DE Wisdom 45

Chapter 3
EXPLORING DIGITAL LEARNING TRENDS 49
    Online Learning 50
    Motivational Innovations 55
    What Changed? 63
    What Would You Do? 65
    DE Wisdom 66

Chapter 4
OVERCOMING THE BARRIERS 69
    Diversity and Digital Equity 70
    Assumptions About Teacher Tech Knowledge 74
    Quality Teaching 75

Contents

    Staying Current, Connected, and Curious 80
    What Would You Do? 86
    DE Wisdom 86

Chapter 5
INNOVATIVE APPROACHES TO DIGITAL EQUITY 89
    Traditional Educator Preparation Courses 90
    Non-Traditional Educator Preparation 95
    What Would You Do? 112
    DE Wisdom 112

CONCLUSION 115

REFERENCES 117

INDEX 133

Foreword

Achieving educational equity for K–12 students is both a longstanding challenge and essential, ongoing goal. Far too often, children from disadvantaged backgrounds attend schools that lag behind those of their more advantaged peers, with fewer academic resources and extracurricular activities, dilapidated facilities, and less experienced and skilled teachers. The increasing adoption of education technologies and growing number of jobs requiring digital literacy skills—including many that have traditionally been performed by workers without a four-year college degree—have introduced a new challenge for educators to address: digital equity.

It's easy for many college-educated professionals to forget that high-speed internet access and computers are still unavailable and/or unaffordable for a significant segment of the population, including many families with school-aged children. According to 2016 U.S. Census statistics, only 81% of households have a broadband internet connection, and only 77% own a home computer (defined as a desktop or laptop).[1] Low-income families, people of color, and rural residents are the most likely to report not having access to high-speed internet and/or a computer at home. In the same vein, not all schools have equal access to technology resources and educators trained to leverage them effectively. In the education context, digital equity includes:

- Access to devices and high-speed internet connectivity, both at school and at home

- Skilled, digitally literate teachers comfortable incorporating technology in instruction

- Educational experiences involving technology that leverage higher-order thinking skills in culturally responsive ways

Unfortunately, the U.S. educational system does not provide equal access to meaningful learning environments powered by technology. Far too often, the classroom technology experiences of lower-income students—especially children of color—are limited to passive technology use, involving activities which do not require active student interaction. Examples include filling out digitized worksheets, "drill and kill"-type activities, and consuming content created by others. Students from more affluent backgrounds are far more likely to use technology in ways that leverage higher-order thinking skills: collaborating with peers, interacting with experts, creating digital content such as videos or blogs, and coding. Often, technology can serve to exacerbate other education inequities instead of narrowing them.

According the U.S. Department of Education, "All teacher preparation programs must prepare graduates to effectively select, evaluate, and use appropriate technologies and resources to create experiences that advance student engagement and learning."[2] This includes the knowledge, skills, and experience needed to leverage technology equitably within a local context. However, too often education technology use in teacher preparation programs is limited to a standalone course that focuses on specific tools, rather than being embedded throughout the curriculum. Digital equity considerations are an afterthought, if even discussed at all.

Fortunately, forward-thinking teacher preparation programs are changing this paradigm, providing preservice teachers with opportunities to use technology in ways that allow for active engagement, through creation, production, and problem solving. Professors are considering the digital equity challenges faced by their own learners and modeling the problem-solving mindset teachers will need in the field. The authors of this book lay out the case for embedding digital equity throughout teacher preparation programs and provide practical, actionable strategies for doing so. By taking a proactive, thoughtful approach to education technology implementation and embedding digital equity considerations throughout, teacher education programs will better prepare educators for the challenges they will face in the field.

There is no "silver bullet" solution that addresses the digital equity challenges of our nation's classrooms. Digital equity, like other educational equity issues, is a complex and multi-faceted problem. However, empowering preservice teachers with the experiences, skills, and mindsets needed to face these challenges is an important step in the right direction.

—Susan M. Bearden
Chief Innovation Officer, CoSN
Author, *Digital Citizenship: A Community Based Approach*

---

[1] Computer and Internet Use in the United States: 2016.
[2] Advancing Educational Technology in Teacher Preparation: Policy Brief

INTRODUCTION

Like us, you may have noticed an increase in the use of the term *digital equity* across educational spaces and places. Defined by researchers as "equal access and opportunity to digital tools, resources, and services to increase digital knowledge, awareness, and skills" (Davis, Fuller, Jackson, Pittman, & Sweet, 2007, p.1), digital equity is a hot discussion topic, but sparse actions are being taken. Sharing a desire to encourage more meaningful efforts to narrow the digital equity gap, we three authors connected through our work with ISTE from across the country. We realized that this work cannot happen in isolation. It must begin in traditional educator preparation programs (also known as teacher preparation programs) before teachers are credentialed and step into their own K–12 classrooms, and it must continue even through non-traditional educator preparation programs that support current classroom teachers. How can we address digital inequities while acknowledging how the growing ubiquity of technology continues to change learning landscapes?

Digital equity is not a new concept, yet the body of literature accessible to a broad audience of educators and decision-makers in teacher preparation on this subject is limited to content that addresses software, hardware, and the digital divide in K–12 environments. Although an abundance of literature exists examining the knowledge and attitudes about technology (Ertmer & Ottenbreit-Leftwich, 2010; Overbay et al., 2010), little evidence has been documented that demonstrates how educator preparation programs are preparing preservice teachers, or supporting current teachers through professional development, to address digital inequities. According to the Office of Educational Technology, K–12 schools should be able to rely on educator preparation programs to adequately prepare new teachers for using technology effectively (U.S. Department of

Education, 2016). This should also include preparing teachers to enact digital equity in their own K–12 classrooms.

For this reason, we offer *Closing the Gap: Digital Equity Strategies for Teacher Prep Programs*, which is the first book in our Digital Equity Series and focuses on closing the digital gap in traditional and non-traditional educator preparation programs. As you read and process the material in these two books, we hope that you will follow and join in the conversation on Twitter using #DigEquityBook.

What's in This Book

In this book, we'd like to encourage you to consider educator preparation programs (also called teacher preparation programs) as you further develop an understanding of critical digital equity issues in order to address challenges with a solution mindset and prepare teachers for their equity-centered classrooms. The ideas put forth in the chapters that follow will challenge you to create spaces and opportunities in educator preparation environments for preservice teachers to apply *digital equity practices*. Digital equity, as noted by Resta and Laferrière (2015), "refers to the social-justice goal of ensuring everyone has equal access to technology tools, computers and the internet, as well as the knowledge and skills to use these resources to enhance their personal lives" (p. 744). Faculty who actively model these practices allow preservice teachers an opportunity to witness effective ways to address digital inequities.

To help teacher education professors and educators working with preservice teachers achieve this goal, Chapter 1 will address the main issues in K–12 that contribute to great concerns pertaining to digital equity in educator preparation programs. A brief overview of technology and digital equity in educator preparation programs through the years will follow in

Chapters 2 and 3, then Chapters 4 and 5 will look to the future, suggesting approaches for overcoming digital equity barriers and enhancing teacher preparedness. Although it may be tempting to skip ahead to the action steps offered in the later chapters, we encourage you read through the historical information to draw inspiration and learn from the shifts that have occurred through the years. The chapters in this book will include the following:

- Current research related to the various facets of digital equity
- Highlights of some current approaches to addressing digital inequities
- Recommendations for moving beyond "naming" a problem as a digital inequity into practical recommendations for closing the digital equity gap
- "DE Stories" or "Open Letters to Teacher Educators" sidebars that feature authentic digital equity stories of successes, lessons from the field, and insights gained from teaching experiences to offer practical examples of how to provide equitable learning opportunities and environments for all students
- "What Would You Do?" sections to encourage innovative thinking about how you may either begin or continue to address digital inequities in teacher preparation
- "DE Wisdom" sections that offer key takeaways of digital equity wisdom as motivation for your own continued efforts enacting digital equity

Throughout the book, you will also see references to the ISTE Standards for Educators (International Society for Technology in Education [ISTE], 2017b). The ISTE Standards are a framework designed to support students, educators, administrators, coaches, and computer science educators with the purposeful integration of technology. Although the ISTE Standards are not the primary focus of the book, it is imperative that consideration be given to the effective inclusion of each strand of the Standards for Educators in teacher preparation courses as well as professional development offerings and non-traditional educator preparation programs.

Who This Book Is For

This book is designed to reflect the contributions of teacher education professors and K–12 leaders seeking to ensure digital equity is achieved on campuses, in classrooms, and throughout educator (or teacher) preparation programs. We will offer historical, philosophical, and practical insights while exploring challenges and solutions unique to educator preparation programs. This book will hopefully inform faculty efforts in pursuit of digital equity within educator preparation programs and their future research. When we discuss educator preparation programs, we are not only including formal credentialing programs at universities, but also are referring to professional development opportunities, educational conferences, unconferences, and informal learning spaces. School administrators, educational technology leaders, current educators, and other stakeholders may also benefit from the positive examples and recommendations offered in this book, especially those related to shifting toward digital equity in teacher preparation as this ultimately influences K–12 campuses as well.

If you are ready to join the pursuit for digital equity, you've selected the right book!

Best,

Nicol, Sarah, and Regina

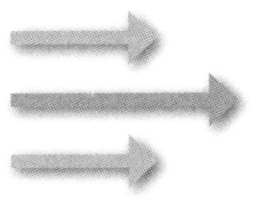

Chapter 1

CRITICAL DIGITAL EQUITY ISSUES

In simple terms, digital equity means all students have adequate access to information and communications technologies for learning and for preparing for the future—regardless of socioeconomic status, physical disability, language, race, gender, or any other characteristics that have been linked with unequal treatment.

—Solomon, 2002

Pause for a moment and think about your technological journey to becoming a teacher. For some of you, learning the nuances of a calculator in your mathematics methods class was the extent of your technology instruction. Those of you who have more recent teacher preparation experiences likely have stories about using some of the latest ed tech tools throughout your programs.

Now imagine facilitating a course that supports students using the latest ed tech tools in a classroom with poor connectivity or a limited number of devices while the school or university across town has 1:1 devices and high-speed internet access. Or, how about two sections of the same course with one taught using no technology to access information or for communication, while the other is facilitated with a connected classroom approach in which students share, communicate, and collaborate with other students globally through blogs and various forms of social media? These scenarios of digital inequity are all too familiar for teachers in K–12 classrooms, as well as for faculty or ed tech leaders leading teacher preparation courses.

According to Culatta (in Molnar, 2014), educator preparation programs are in need of help. The poor design and delivery of digital learning experiences in educator preparation can create barriers for future and current teachers, which in turn leads to more difficulties in the classroom. Poorly designed and delivered digital learning experiences often merely try to replicate a paper-based lesson with internet clicks. For example, a professor replacing a paper worksheet with an online version of the same activity does not demonstrate the full potential of digital learning experience to preservice teachers in his class. Inevitably K–12 students suffer from the lack of teacher preparedness for dealing with digital inequities, because preservice teachers potentially reenact the same digital inequities they experience in their preparation programs.

In education, we can aim to ensure students have equal access to devices, software, and the internet, and yet if they do not always have educators trained in the use of technology to help them navigate those tools, then we cannot achieve true digital equity. Educator preparation, both traditional and nontraditional, would be remiss if it did not support the integration of digital equity lessons and practices throughout its programs. In this chapter, we will offer a brief overview of digital equity in K–12 education today, then define the critical issues and disparities in teacher preparation that can sabotage efforts towards digital equity. Our goal in this chapter is to address the main issues in K–12 that contribute to great concerns pertaining to digital equity in educator preparation programs.

Brief Overview of Digital Equity in K–12 Today

To achieve digital equity, something has to change in education. As James Ford, 2014 North Carolina Teacher of the Year, explained at The Friday Institute's 2016 Redefining Equity for Digital-Age Learning Convening:

> If we are to adequately fulfill our mission as educators, we need to adopt a new way of doing things. We need to recognize the conditions, do away with the constraints of convention and embrace a new model of education. We need to radically rethink what school looks like in 21st-century America, and apply an equity lens to our work. (Ford in Smart & Corn, 2017, p. 8)

Before we can embrace that new model of education, however, we need to understand our students' current conditions as well as how they're using technology at home and at school. Rideout and Katz (2016) conducted a survey to see what exactly our students from ages 10 to 13 were doing with technology once they got home from school. The researchers found the following:

> Among 10- to 13-year-olds who use computers or the Internet, eight in ten (81%) do so to do homework, and four in ten (44%) to write stories or blogs. Many also use the Internet to connect with teachers (40%) and

other students (46%) about school projects. Among 6- to 13-year-olds, 81% play educational games and use the Internet to look up things that they are interested in. Seven in ten (70%) use computers or the Internet to do something creative, such as make their own art or music. (p. 34)

Rideout and Katz also found that while 94% of families surveyed reported having internet access, the type of access varied as "one quarter (23%) of families below the median income level and one third (33%) of those below the poverty level rely on mobile-only internet access. And many experience interruptions to their internet service or constrained access to digital devices" (p. 5).

Other challenges included slow access, sharing devices between multiple members of the household, and having service cut off because of nonpayment. The researchers found no significant differences in students' home internet use, regardless of their level of access. There is also no indication that students lack an interest in using devices at home, when accessible (Figure 1.1).

From this data, we see that students at various income levels are generally motivated to use the internet for similar reasons at home; depending on the resources available at home, however, the outcomes may be very different. Consider, for example, an assignment such as watching a flipped video lesson. A student with high-speed access and a dedicated device would likely have few to no barriers in completing this assignment, at least as far as access is concerned. But, what about students with mobile-only access? Those who have to share a device with multiple family members? Those with frequent service interruptions? For a student facing such challenges, a flipped video homework assignment may not be as simple and straightforward to complete.

Figure 1.1 K–12 students, such as this second grader, are motivated to use technology at home.

As an education community, we need to prepare teachers to examine issues such as these, to ensure that their instructional design is equitable, and not inadvertently widening such gaps in digital equity. As educators, we do what is necessary to meet the needs of every student, but as eloquently stated by Jennie Magiera (Google Innovator Program, 2017), we must also be intentional in meeting the needs of *each* student. With this in mind, next we will briefly examine the state of affairs in digital equity in K–12 education to identify which conventions work and which need to be "radically rethought."

Opportunity Gaps

Learners with diverse learning needs—and particularly those living in poverty—experience educational "opportunity gaps." Although schools are making progress towards providing equal access to technology, access alone does not guarantee that students of all backgrounds and abilities have equal opportunities to learn. The promise of technology depends on creating ongoing opportunities for educators to continuously improve their instructional practice (Alliance for Excellence in Education, 2016). By now you know that digital equity goes far beyond simply having access to technology (which is not as "simple" as it may seem either). A variety of factors play a role in true digital equity, not the least of which is access to high-quality learning experiences that are facilitated by teachers.

As the Alliance for Excellence in Education pointed out, students living in poverty often do not have access to these learning experiences and high-quality teachers. The U.S. Department of Education Office for Civil Rights (2014) came to a similar conclusion, stating "Schools serving the most black and Latino students are 1.5 times more likely to employ teachers who are newest to the profession (who are on average less effective than their more experienced colleagues)" (p. 4). Many of these educators are recent graduates of educator preparation programs, which is why it is important to make sure that the education community is doing all that we can to help preservice teachers meet the needs of each child they serve. Often, when the term *gap* is used in an educational context, data are presented solely in the context of achievement gaps. Commonly absent from the discussion are the gaps in the preparation of teachers and high-quality instruction using emerging technologies.

A Silver Lining

Despite the challenges in K–12 education regarding digital equity, colleges and universities have the distinct advantage of being able to foster best practices at a critical time for preservice teachers. Those teaching the next generation of teachers are in a position to help education students leverage the power of instructional technology in authentic and meaningful ways, so they, in turn, can make learning stick in their own future classrooms! As the U.S. Department of Education Office of Educational Technology (2017) stated:

> Technology can be a powerful tool for transforming learning. It can help affirm and advance relationships between educators and students, reinvent our approaches to learning and collaboration, shrink long-standing equity and accessibility gaps, and adapt learning experiences to meet the needs of all learners. (p. 3)

Each student is different and comes to the classroom with unique skills, challenges, and needs. By learning and applying strategies such as differentiation and personalization in combination with technology, educators are able to help meet students where they are. In a 2017 nationwide Speak Up Survey, Project Tomorrow researchers found that "teachers in blended learning classrooms are setting a new bar for transforming learning using technology. For example, 68% report that with the use of technology in their classroom they are better able to differentiate instruction for their students" (Project Tomorrow, 2017). With such a large percentage of in-service teachers reporting the effectiveness of technology on student achievement, teacher candidates must be prepared to follow and advance this trend.

How might colleges and universities accomplish this? Project Tomorrow also revealed that when teachers experience technology as a learner, they also "have stronger valuations on the role of technology within learning, and higher aspirations for leveraging technology to support transformed learning environments" (2017). In sum, modeling best practices, many of which we will discuss throughout the course of this book, is one of the best approaches to helping preservice teachers.

CHAPTER 1 • Critical Digital Equity Issues

 DE Stories

Matthew Hiefield's Perspective

Student access to robust digital tools and curriculum is key to their success as 21st-century citizens. Yet many students from economically disadvantaged families have limited access to these tools outside of school. In addition, nearly two-thirds of school system leaders do not have any strategies for providing off-campus connectivity to students (CoSN Infrastructure Survey, 2016). Closing this opportunity gap requires a variety of strategies involving schools and communities working together. In Beaverton, Oregon, the Beaverton School District began addressing this challenge by starting a monthly "Digital Equity Brown Bag Lunch" discussion group. Beaverton cast a wide net and extended invitations to people from different schools and different departments. Really, the main requirement was to show up and to bring focused energy to this issue. The monthly meetings didn't jump directly to solutions. Rather, Beaverton gathered data and wondered what more information was needed. Different people brought a wide variety of perspectives, and that led to solutions that might not have been thought of in a small, similar group.

After a year of study, some initial action items emerged:

- **Extended library hours.** One outcome to support students was to extend library hours beyond the normal school day. Although helpful to many, it didn't help students who worked and/or took the bus home. This year, our district also provided extended-activity buses late in the day, so that students could stay after school to access school

Wi-Fi and other digital resources, while still taking a bus home.

- **Hotspot pilot project.** In 2016–17, Beaverton piloted a hotspot project in one of our high schools to increase access outside of school hours. We learned quickly that demand far exceeded the available supply. The lessons from this pilot project informed our decision to apply for hotspot grants.

- **Wi-Fi access maps.** Some of our schools in highly impacted areas contacted local businesses to see if they would be willing to allow students to use their Wi-Fi after school hours. This proactive approach helped build community support and engage others with our access challenges. To spread the word to students and their parents, Beaverton produced English and Spanish brochures that mapped and detailed available Wi-Fi sites. You can see the English version at goo.gl/NFGYZF.

- **Sprint 1Million Project grant.** In the Spring of 2017, our Digital Equity Team applied for a Sprint 1Million Project grant to fund additional hotspots. Beaverton was fortunate enough to be chosen, and the following fall we deployed hotspots in all of our high schools. In planning to distribute the hotspots, we emphasized making the experience an affirming process for students and not one that would make students feel uncomfortable when receiving their hotspot. The project will continue through 2025. To view one of our rollouts, visit vimeo.com/234439613.

- **Latino parent tech nights.** Beaverton began hosting technology nights to educate Latino parents on how to support their students with technology. Many of these parents had not used a computer before, so educating them about digital citizenship, expectations, and parenting helped to show them the value of being connected and ready for the future. We also used this time to discuss broadband access challenges as well as parenting strategies with technology. (For more details, visit goo.gl/E9WfcR.)

- **Kajeet hotspot project.** To meet the needs of middle and elementary schools, Beaverton is piloting a hotspot project with Kajeet (kajeet.net) at three of our schools with the highest needs. We are working closely to gather data on the impact of using the hotspots on teaching and learning at these schools in order to inform future adoptions.

- **Summer access reading project for prekindergarten students.** Beaverton is also piloting a project at Title One schools in which three incoming students each will be provided with an iPad during the summer. They and their families can use the iPad to explore learning opportunities, local community activities, and digital books to be read aloud. Our goal is for this early connection to be a welcoming and informative support to their upcoming entrance into kindergarten. Families with internet access issues will be supported with hotspots. We are hoping that fostering literacy skills before our students reach our schools will provide both immediate and long-term benefits.

- **District Survey.** We include questions about home internet connectivity in our district-wide student survey and also administer, site-specific, connectivity surveys at high poverty schools.

> Connectivity issues are nuanced and complex, and no single initiative can solve everything. However, consistently focusing on these issues and brainstorming solutions are fundamental in fostering digital equity for all of our students. It is tempting to try to solve connectivity issues with one meeting and one program. The real key is to organize a dedicated group of people who are willing to focus on digital equity challenges on a continuous basis and who are willing to adapt and explore new ideas as schools and technology evolve.

Defining the Critical Issues in Teacher Preparation

When considering the critical issues in teacher preparation pertaining to digital equity, it helps to have a baseline definition of digital equity itself, such as:

> Digital equity is defined as equal access and opportunity to digital tools, resources, and services, to increase digital knowledge, awareness, and skills. When considering the role of technology in development of the 21st-century learner, digital equity is more than a comparable delivery of goods and services, but fair distribution based on students' needs. (Davis et al., 2007, p. 9)

Although the definition was originally describing K–12 classrooms, it can also be used to understand the digital inequities encountered by future teachers, current teachers, and faculty in teacher preparation. According to the National Education Technology Plan (NETP; U.S. Department of Education, 2017), equity issues also include barriers faced relating to race, ethnicity, national origin, gender identity, disability, English language ability, religion, socioeconomic status, or geographic location. These issues can be seen across educational spaces (kindergarten through higher education) and are not limited to non-technology related contexts. In other words, the issue of equity is not isolated to face-to-face human interactions or social exchanges. Hence the term *digital equity* not only refers to the distribution of devices and issues related to access, but also includes technology practices and beliefs.

Similar to in K–12 classrooms, we specifically see digital inequities in teacher preparation through decision-making related to teaching, access to the internet and devices, and implementation plans (or tool selections). In order to achieve digital equity in education, access to educational opportunities must be increased with a focus on closing achievement and digital equity gaps by eliminating such barriers as poor-quality teaching, lack of access, and poor choices related to tool selections or implementations in teacher preparation. Although this chapter will include a few actionable tips, you will find more in-depth solutions and examples in Chapter 4.

Quality Teaching

According to Culatta (in Cortez, 2017), the future of technology in education is about using technology as a tool to accelerate innovation and less about the simple delivery of instruction using technology. The U.S. Department of Education echoed this belief, saying:

> On its own, access to connectivity and devices does not guarantee access to engaging educational experiences or a quality education. Without thoughtful intervention and attention to the way technology is used for learning, the digital use divide could grow even as access to technology in schools increases. (2016, p. 20)

Similar to in K–12 classrooms, there are moments in formal and informal educator preparation programs when a professor must choose whether to deliver a lesson using direct instruction or a learner-centered experience. A professor must decide whether to teach a lesson on collaboration, for example, in lecture format (direct instruction) or immerse preservice teachers in a learner-centered experience that calls for them to actually collaborate with peers locally and/or globally. In addition to deciding how they will deliver instruction, professors must also determine which technologies may work best for any given activity when not all preservice teachers may have access to their own devices or internet at home. Situations such as this are not uncommon for preservice teachers, especially for those on tight budgets or those living in rural communities. So for the same collaboration lesson mentioned above, the professor may have decided to facilitate a local in-class collaborative activity with preservice teachers in groups of four, using Google Docs, in which each group member contributes to the shared document. The professor may hope for a 1:1 environment in which each preservice teacher has a dedicated device; however, equity-minded professors should also prepare for the possibility that not everyone has a device to bring to class. In situations such as this, shifting to a 4:1 environment in which one group of four contributes as a whole to a Google Docs document shared with another group of four may serve as an alternative activity. Regardless of the grouping strategy, the activity puts the technology in the learner's hands. The professor is offering preservice teachers a chance to explore a pedagogical approach made possible by technology (ISTE Standard for Educators 1a, 2017b). Instructors who push beyond direct instruction and consider the elements of quality teaching, such as thoughtful interventions and attention to the way technology is used for learning, are well on their way in the pursuit of digital equity.

In Chapter 2, we will further discuss how to establish and build upon a strong teaching foundation in pursuit of digital equity, as well as how the ISTE Standards for Educators can be leveraged in educator preparation programs to further support the development of quality teaching. For now, let us take a closer look at two more critical digital equity issues in teacher preparation: access to internet and devices, as well as the implementation of technology plans.

 Reflect & Consider

Quality Teaching

- Do you assess whether your preservice teachers have technology experience or prior exposure to use of educational technology in their own K–12 background experience? How might this data inform your use of technology in your classes?

- Do you assign projects that allow preservice teachers to model their design and delivery of digital learning experiences and offer rich feedback to support their development of digital equity strategies?

- How often do you check in with your preservice teachers to inquire about what tech-rich teaching they are observing at school sites during their fieldwork or student-teaching experiences?

Access to Internet and Devices

It can be very difficult for educators to manage learning experiences and activities when they have to support connectivity issues. When faced with challenges, educators may revert to activities of the lowest common denominator at the expense of a more effective learning experience (U.S. Department of Education, 2017, p. 76.). In the collaborative activity described in the previous section, a professor who is overly concerned about potential connectivity issues may forego using technology and instead present an offline group lesson. Connectivity concerns may influence a professor to forego any out-of-class global collaborations between his or her preservice teachers and others around the world, as well. Although an offline group activity may offer preservice teachers a chance to explore how to facilitate collaborative group activities in their own classrooms, their professor has potentially missed an opportunity to use collaborative tools to expand preservice teachers authentic learning experiences through engagement with other classmates locally and/or globally (ISTE Standards for Educators 4c, 2017b). Other, better solutions are available. Rather than pass opportunities by because of potential issues, we need to educate ourselves on the way our students do have access to and are using technology outside the classroom. Armed with more information, we can make better choices. Through better decision-making, we can also support preservice students who live and may eventually work in rural communities and may be faced with their own connectivity concerns in their future classrooms. Although a few strategies related to rural schools will be offered in this chapter, solutions for this topic will be explored further in the next book in this series, *Closing the Gap: Digital Equity Strategies for the K–12 Classroom*.

When the term *digital divide* was introduced in the early 1990s, the main focus was access to computers. According to the U.S. Department of Education (2016), by the late '90s the National Telecommunications and Information Administration (NTIA) identified a digital divide issue and named the United States a nation divided. Findings from the National Center for Education Statistics (U.S. Department of Education, 2016) survey support this assertion and indicate a pattern of significant differences in household computer access through the years. Table 1.1 below shows the disparities by race/ethnicity in access to devices (desktops, laptops, tablets, and smartphones), as reported by NCES from the late '90s to 2015.

Table 1.1 NCES Data on Computer and Device Access

Year	Race/Ethnicity	Computer Access (percent using by household)
1997	White	36.9%
	Black	15.6%
	Hispanic	14.5%
2003	White	63.7%
	Black	40.8%
	Hispanic	35.8%

Year	Race/Ethnicity	Device Access (percent of population ages 3 to 18)
2010	White	92.4%
	Black	72.8%
	Hispanic	74.2%
2013	White	95.9%
	Black	87.1%
	Hispanic	87.2%
2015	White	97%
	Black	90.2%
	Hispanic	90.7%

Access to computers in public schools over the years has mirrored the disparities shown in Table 1.1 and further confirms the historical digital divide by race/ethnicity. Although programs such as E-Rate (https://www.fcc.gov/general/e-rate-schools-libraries-usf-program) have emerged to provide devices and internet access at a lower cost, criticism persists with regards to the device and hotspot disbursement methods as well as the level of use of computers by teachers on school campuses. Increasing the number of devices on campuses is only the first step to closing digital equity gaps, which is why increasing the numbers of hotspots distributed for use at home has been of great benefit to students from low-income income families and those living in rural areas that lack affordable internet access.

Professors in educator preparation programs would be remiss not to discuss with preservice teachers the potential obstacles they may face in rural districts, as well as possible solutions. Additionally, professors should consider the possibility that their own students (preservice teachers) may dwell in rural areas. For example, a professor may discover that even with hotspots internet access is sparse for preservice teachers at home due to poor cellular coverage. This could potentially cause frustration when completing online activities from home, such as collaborative research projects using Google Docs or discussion board prompts. Preservice teachers are likely to assume that if their future district is in a rural area that their future K–12 students will experience similar frustrations. Instead of leaving preservice teachers to their own assumptions, their professors can engage them in solution-based dialogues about how to enact digital equity when the most pernicious obstacles present themselves. Here are a few thoughts to consider when preparing preservice teachers for rural areas:

- After teaching with a new technology tool, how often do you ask "what would you do if you were teaching a lesson using this same tool and the Wi-Fi goes down in your classroom?"
- Do you assign projects that allow preservice teachers to model their design and delivery of offline learning experiences that might allow for a seamless transfer of knowledge when learning with technology?

- How often do you discuss with preservice teachers the importance of how to clearly communicate the benefits of using technology to parents and key stakeholders who might be skeptical of the role of technology in society?

Professors in educator preparation programs must also prepare future teachers to effectively choose and use emerging technologies in their classrooms. The disparities in frequency and choice of technology use by teachers raise greater concern as we begin to see greater numbers of devices (desktops, laptops, tablets, and smartphones) in homes and on K–12 campuses (Figure 1.2). Based upon NCES data, the pursuit of digital equity in K–12 and in teacher preparation must persist, yet the focus should not be solely on physical access to computers but rather on how (and how often) technology is used in teaching and learning.

Figure 1.2 K–12 districts have shifted to 1:1 devices on campuses, which calls for a better preparation of teachers

Efforts to decrease the digital divide address *what* we will do; however, the response to the question of "how?" naturally shifts

Critical Digital Equity Issues • CHAPTER 1

the conversation from the use of the term digital divide to digital equity. These terms are often used interchangeably and frequently misunderstood. In one study by Voithofer (2009), preservice teachers were asked to define the digital divide and more than half of the participants either left the response blank or indicated they were not familiar with the digital divide. In contrast to the term *digital divide*, the term *digital equity* is most certainly about more than physical access to technology (Warschauer, 2007). Faculty supporting preservice teachers with understanding the differences is essential and sets the stage for a greater focus on how to address digital inequities their future students are likely to encounter.

In addition to supporting preservice teachers with their understanding of the digital divide, it is also imperative that professors take note of their students' device access. Just as in K–12 classrooms, limited access, connectivity issues, and disparities in teaching choices can impede the pursuit of digital equity in educator preparation programs, whether they're formal credentialing programs at universities, professional development opportunities, educational conferences, or informal learning spaces (e.g., EdCamps, CoffeeEdu, EduMatch, Unconferences). In cases when not everyone has their own device, some universities offer options for checking out laptops and tablets, similar to borrowing library books. Often times, preservice teachers may opt to use their smartphones over laptops. Because mobile devices have difficulties displaying certain websites and apps, faculty may need to reevaluate tool selections and/or consider innovative grouping strategies with preservice teachers working in groups with one shared device (e.g., 3:1 or 4:1).

Reflect & Consider

Access and Devices

- Do you assess whether your preservice teachers have access to devices prior to the start of class? How does this information inform your pedagogical practice and collaborative grouping strategies?

- Do you inquire with your preservice teachers about their connectivity at home, or availability to work from campus, prior to assigning homework?

- Do you prepare options for completing in-class tasks in the event that devices are limited or the Wi-Fi goes down?

▲ ▲ ▲

Implementation Plans

Similar to K–12 settings, the adoption of new technologies in educator preparation programs requires sustainable implementation plans and must begin with a clear purpose and rationale for the new tool. As Clark pointed out:

> Teacher educators must be aware of emerging trends in technology; at the same time, they must understand the potential, as well as the pitfalls, of purchasing, adopting, and using a wide array of tools. (Clark, 2009, in Dieker, Kennedy, Smith, Vasquez, Rock, & Thomas, 2014)

Teachers' learning needs related to technology integration have been historically documented as insufficiently supported through formal learning opportunities, and the formal and informal support they receive is imperative throughout the implementation process (Jones & Dexter, 2018). Common adoption practices often include deficit uses of technology, such as technology viewed as an additional subject, implementation of technologies for drill and practice, and the simple replacement of existing paper-based assignments with technological versions (Warschauer & Matuchniak, 2010). Results from the 2017 Speak Up Survey indicated that 67% of the technology leader participants agreed that the greatest challenge they face in implementing digital learning or expanding technology use is motivating teachers to change their traditional instructional practices to use technology more meaningfully with students (Project Tomorrow, 2017). In educator preparation programs, implementation plans can occur at the school or college level and even at the course level. Adoptions of new technologies at the course level are easier for faculty and allow them to fairly assess a tool and address any digital inequities.

Reflect & Consider

Implementation Plans

- How might you contribute to the sustainability of new technology implementations at your university?

- As you implement new technologies into your courses, are you pushing past the simple replacement of pen and pencil with innovative technologies?

- What factors are you considering when assessing the tools you will introduce in your courses?

- How are you encouraging preservice teachers to reflect on their own pedagogical choices when implementing innovative technologies?

An Open Letter
from Dr. Nirmla Flores

Dear Teacher Educator,

As I write this brief letter, I cannot help but admire your effort in dedicating time to amass intellectual capital, especially in the realm of technology. The fact that you are reading this book is reassuring, as it shows your desire to learn and expand your technological skills, insights, techniques, and strategies to become more adept at what you do. I truly applaud you for this! Not only are you doing this for yourself, but you are also benefiting the lives of your students (present and future) as you impart your knowledge and wisdom to them.

I am always energized in seeing such amazing teacher educators who have the passion to learn, to grow, and to share. As a junior faculty member, everything seems to be fresh and new in my eyes. Even though I have been in the field of education for almost twenty years, mostly as an elementary school teacher, I constantly see young and seasoned educators alike, having the burning desire to develop themselves by being open to different possibilities. As a self-proclaimed "newbie," I too am constantly learning the ropes on how to navigate through the maze of higher education (i.e. teaching, scholarship, and service). Through it all, I see the many challenges and successes that life in the university can bring.

One major challenge that I have discovered throughout this process of navigating is my limitations in technology. I feel that I am always playing catch-up: As soon as I figure out a certain technological tool, program, or device, a newer version or a much more sophisticated adaptation has surfaced. I find it hard to keep up with all the new technologies that come my way and then seemingly pass by me at the speed of light. Indeed, technology is constantly and rapidly changing before my very eyes!

The great news is that my university has opportunities, such as educational technology grants, which have truly opened my eyes to new possibilities. One recently funded project called "Achieving Digital Equity in Teacher Education" was an exhilarating journey with an indescribable excitement from start to finish. Much like being on a roller-coaster ride, there were some moments where I felt suspended in slow transitions, knowing that I was about to accelerate in no time. Sure enough, the accelerations came up so fast that I thought I could not get a good grip on what I needed to hold on to. Thankfully, in the "Achieving Digital Equity in Teacher Education" technology grant project, I worked with two outstanding faculty members whose talent is exceeded only by their dedication.

What a privilege it has been to be able to join a team of faculty members whom I can rely on, especially in a massive project which I feel ill equipped to handle. The huge "aha" moment and the secret that I have come to realize is to surround myself with brilliant colleagues who share the same sentiments, passion, and yearning as I do. No doubt, collaboration in technology is one of the keys to success in life-long learning! I encourage you to team up and team up often with your kindred spirits in education. Go as far as you can see in the technology world, side by side with others. And I can almost guarantee, you will see farther!

What Would You Do?

Imagine a well-planned learner-centered lesson in which a professor gives a group of future educators a collaborative in-class task to design a fictitious school and to create an accompanying website. When everyone arrives for class, the professor notices that a handful of students do not have their own devices and of all days for poor Wi-Fi connectivity, today is it. *What would you do?*

You might consider taking five to ten minutes to ask students to brainstorm and troubleshoot how they would adjust the lesson if they were leading their own classrooms. They will likely create a list of innovative ideas; however, if they are stumped you might push forward and simply have them ask their class partners to share devices (preferably smartphones on unlimited data plans).

Tell us what you would do on Twitter using #DigEquityBook.

DE Wisdom

> Students learn in different ways, and often culture does play a role. As a matter of fact, the intersection of all of our cultures plays a huge role in our learning. Respect and embrace that. Each learner is different and brings with them a wealth of cultural capital. Don't be afraid to try new strategies with your students, but make sure you are also respecting them as individuals.
>
> —*Dr. Sarah Thomas*

Similar to K–12 students, preservice teachers learn in different ways and their cultural backgrounds also play a role in their learning. Exposing preservice teachers to new tech-rich strategies through demonstration lessons, modeling digital equity, and acknowledging and respecting their learning differences prepares them to do the same with their future students. We challenge teacher educators to not only consider the varying levels of technology knowledge of those who they teach, but also to consider what role their learners' backgrounds may play in how they approach enacting digital equity.

If you have tips to share on this topic, please share them on Twitter using #DigEquityBook.

Chapter 2

LEVERAGING MOBILE AND SOCIAL TRENDS

> The use of technology has exploded throughout every avenue of society and education; teacher education is no exception ... teacher preparation programs are held accountable for their graduates and how well they improve achievement outcomes for K–12 students.
>
> — Clark, 2009, in Dieker, Kennedy, Smith, Vasquez, Rock, & Thomas, 2014

For preservice teachers to be prepared to enact digital equity and facilitate meaningful learning in their future classrooms, they must receive better support in their technology development today. We must also better prepare tomorrow's educators to facilitate learning opportunities that foster creativity using emerging technologies.

According to the National Education Technology Plan (U.S. Department of Education, 2016), educators should leave teacher preparation with a solid grasp of how to utilize technology to support the learning of K–12 students. That said, "few traditional teacher preparation programs yield teachers who are equipped for success in our complex school environments" (Holt & Garcia, n.d.). Over the years, we have seen progress in teacher preparation, yet the nature of emerging technologies forces us to continue building new experiences for deeper explorations in educator preparation programs so preservice teachers feel better prepared to address digital inequities in today's classrooms. Because of the rapid pace of technology, researching and understanding its past integration into teacher education programs can be challenging. For example, in earlier literature on technology in teacher education, the term *digital equity* was often used to describe simply the access to devices and/or the internet. As noted in the previous chapter, however, digital equity involves so much more.

Chapter 1 provided some basic context surrounding the topic of digital equity in educator preparation programs, specifically through the lens of K–12 education. In this chapter and in Chapter 3, we will continue to build upon that foundation by exploring how technology fits into preservice teacher education and examine how shifts in technology trends can influence the ways in which preservice teachers and those who educate them enact digital equity.

The Horizon Report

When making sense of the shift regarding technology in educator preparation programs, the New Media Consortium's Horizon Report is a good place to begin, because the innovations addressed were projected to be the current best practices in the landscape of higher education. Table 2.1 summarizes the technology adoption predictions identified by the New Media Consortium (NMC) through the years, with notes for clarity indicated by brackets. (You can find the full reports at www.nmc.org/publication-type/horizon-report.)

Table 2.1 NMC Horizon Report Technology Adoption Predictions

Year of Report	Mainstream Use in 1 Year or Less	Mainstream Use in 2 to 3 Years	Mainstream Use in 4 to 5 Years
2004	Learning Objects, Scalable Vector Graphics	Rapid Prototyping, Multimodal Interfaces	Context-Aware Computing, Knowledge Webs
2005	Extended Learning, Ubiquitous Wireless [Connection]	Intelligent Searching, Educational Gaming	Social Networks, Knowledge Webs, Context-Aware Computing/ Augmented Reality
2006	Social Computing, Personal Broadcasting	Phones in Their Pockets [mobile phones], Educational Gaming	Augmented Reality, Enhanced Visualization, Context-Aware Environments and Devices
2007	User-Created Content, Social Networking	Mobile Phones, Virtual Worlds	The New Scholarship [and Emerging Forms of Publication], Massively Multiplayer Educational Gaming
2008	Grassroots Video, Collaboration Webs	Mobile Broadband, Data Mashups	Collective Intelligence, Social Operating Systems
2009	Mobiles, Cloud Computing	Geo-Everything, The Personal Web	Semantic-Aware Operations, Smart Objects

Continued

CHAPTER 2 • Leveraging Mobile and Social Trends

Year of Report	Mainstream Use in 1 Year or Less	Mainstream Use in 2 to 3 Years	Mainstream Use in 4 to 5 Years
2010	Mobile Computing, Open Content	Electronic Books, Simple Augmented Reality	Gesture-Based Computing, Visual Data Analysis
2011	Electronic Books, Mobiles	Augmented Reality, Game-Based Learning	Gesture-Based Computing, Learning Analytics
2012	Mobile Apps, Tablet Computing	Game-Based Learning, Learning Analytics	Gesture-Based Computing, Internet of Things
2013	Massively Open Online Courses, Tablet Computing	Games and Gamification, Learning Analytics	3D Printing, Wearable Technology
2014	Flipped Classroom, Learning Analytics	3D Printing, Games and Gamification	Quantified Self, Virtual Assistants
2015	Bring Your Own Device (BYOD), Flipped Classroom	Makerspaces, Wearable Technology	Adaptive Learning Technologies, The Internet of Things
2016	BYOD, Learning Analytics and Adaptive Learning	Augmented and Virtual Reality, Makerspaces	Affective Computing, Robotics
2017	Adaptive Learning Technologies, Mobile Learning	The Internet of Things, Next Generation LMS [Learning Management Systems]	Artificial Intelligence, Natural User Interfaces
2018 (not yet published at the time of this writing)	Analytics Technologies, Makerspaces	Adaptive Learning Technologies, Artificial Intelligence	Mixed Reality, Robotics

As you can see, three trends stand out for their frequent reoccurrence in various guises:

- Mobile Devices
- Social Media
- Personal Broadcasting

The following sections take a closer look at how each of these broaden the classroom experience, while Chapter 3 will delve deeper into the online learning advances and motivational innovations identified by the NMC.

Mobile Devices

The Horizon Report repeatedly identified mobile phone use as a technology adoption to watch from 2006 onwards, especially after the wide acceptance of smartphones following the debut of the iPhone in 2007. The challenge for educators has been to capitalize on the mobile phones students bring to class, without these devices becoming a distraction. NMC identified "**the phones in their pockets**" as major influencers in higher education by 2009, pointing out how the advent of third-party applications, such as advanced calculator apps, music simulators, and more, opened new possibilities in higher education. Improvements such as increased battery life, better displays, and elevated storage capacities were also noted as potential game-changers in higher education. The proliferation of personal mobile devices opened more avenues for equity. For example, preservice teachers with limited funds became able to participate in activities requiring the internet by using their mobile devices instead of having to purchase a laptop computer.

Tablets and BYOD

In 2012, the Horizon Report projected an increased use of **tablets**, as well. With the ability for users to load apps and content, the tablet was referenced by the NMC as a portable personalized learning environment. With an increase in mobile devices, such as smartphones and tablets, some universities implemented 1:1 programs, but

> Where one-to-one learning is not yet possible, many institutions, including the Community College of Aurora in Colorado, the University of Richmond, and the University of South Carolina, have also made tablets available via check-out systems to students who may not have one, in which students can borrow tablets to do coursework that is specifically designed to be completed with the devices. (2013, p. 17)

The mere fact that professors were assigning work that required use of the tablets spoke to how ingrained they had become in the culture of some universities by 2013. Mobile device use, including the use of tablets and apps, has become compelling enough that schools are rethinking standing policies that prohibited the use of mobile devices in class and some are even implementing "bring your own device" (**BYOD**) programs. BYOD programs or 1:1 initiatives, where each learner possesses an iPad (tablet) or Chromebook continue to grow in popularity.

Digital equity must be carefully considered in programs such as these, however, to ensure that each student has high-quality internet access. Furthermore, while many in-service teachers are open to integrating technology to improve learning outcomes, there are many who are not fully prepared to do so. This can exacerbate digital equity gaps, and further underscores the need for institutions of higher education to prepare preservice teachers with such skills. Additionally, educator preparation programs should partner with districts to support in-service teachers with informal teacher preparation opportunities (such as professional development) to close the digital equity gap between current and future teachers.

Apps and Mobile Learning

While BYOD and tablet use rose through the years, the use of smartphones and **apps** increased as well. In fact, mobile learning reappeared in the Horizon Report in 2017 for the first time since 2012. According to NMC, mobile learning was often equated with extended learning or distance learning and allowed educators to "take advantage of an enriched environment where classroom instruction is supplemented by an interactive component facilitated by technology tools" and "take advantage of tools that students already carry" (p. 6). Extended learning opportunities also made use of tools such as wikis, text messaging, and even blogs, which had been identified as mainstream technology by CNN in 2004.

The mobile learning innovation had far-reaching implications for equity; in many cases learners no longer had to buy expensive single-function devices and could instead download a free or inexpensive application on their phones with the same functionality. It has become common "to use cloud-based apps to manage calendars, rosters, grade books, and communication between school and home" (p. 10). Teaching preservice teachers about such cloud-based applications prepares them to enact digital equity, as their future students' parents who are unable to stop by the classroom will be able to gain more insight into their child's progress, as long as they have internet access available.

By 2012, tablets were taking the higher education world by storm, and institutions of higher education were also beginning to develop their own apps for student use. Furthermore the possibility for anyone to develop an app was a potential contributing factor to the drive behind coding and computer science movements, which we have seen gain traction over the last few years in K–12 education. As K–12 continues to embrace the possibilities of mobile learning, educator preparation programs are increasingly doing the same.

Reflect & Consider

Mobile Trends

- As you consider the mobile trends, are you protecting personal data, your digital identity, and data privacy? ISTE Standard for Educators 3d (ISTE, 2017b) encourages educators to not only engage in these practices, but to model and promote them as well. Doing so raises awareness for preservice teachers as they begin to explore using mobile devices, tablets, and apps in their future classrooms.

- How are you encouraging the innovative use of apps in your classrooms? Are you allowing preservice teachers to select apps to create authentic learning experiences for K–12 students that maximize learning (ISTE Standard for Educators 5b, 2017b)?

- Are you designing learning experiences for your preservice teachers that allow them to create innovative tools, such as apps, to support their acquisition of the necessary skills to be able to offer their future students similar opportunities to innovate?

Social Media

Given the rise of social media in the world at large in 2007, it was not surprising that social networking was forecast for short-term implementation in education by the NMC that year. According to the 2007 Horizon Report authors, "Social networking is already second nature to many students; our challenge is to apply it to education" (p. 12). They identified that building community was the driving force in its popularity and advocated that professors tap into these tools to encourage creativity, create authentic learning experiences, and extend face-to-face opportunities beyond the length of the event itself. Educators were urged to implement social media platforms to help students display work to authentic audiences and make connections to others. By 2008, tools such as Facebook, MySpace, Flickr, and Twitter were among the platforms listed for use. Although some of the major social media players have changed since then, their overall presence has increased. For example, you often now see Twitter hashtags associated with a course or conference.

Perhaps the biggest technology—and social media—story of 2014 was about its use rather than a new innovation. The world continued to see how impactful grassroots video and social media had become to our society when a police officer shot and killed unarmed teenager Michael Brown in Ferguson, Missouri. As people had the power to share their own truths via social media, traditional media gatekeepers were unable to spin and contain a narrative. Citizens shared what was happening up-to-the-minute on Twitter, while most news outlets initially chose not to cover the story.

Similarly, social media networks might empower students to advocate for global change to authentic audiences (Krutka & Carpenter, 2016). By 2016, social media made headlines among youth internationally, ranging from the silly (mannequin challenges) to the serious (organized protests for social justice). Social media potentially allows K–12 students and preservice teachers an opportunity to share their voices and their own truths, so teacher educators would be remiss not to prepare future educators for navigating these spaces with their future students. In their article, *8 Considerations for Social Networks in Classrooms,* Howard and Howard (2014) offered thought-provoking questions for teacher educators and preservice teachers who are contemplating using social media networking sites in their classrooms:

- What is the impact of classroom social networking sites use on your students' motivation?

- Do efforts at incorporating a particular social networking site unfairly disadvantage certain groups due to their lack of proficiency in and/or familiarity with its interface and functionality?

- Do social networking sites encourage inappropriate or negative social behavior amongst students in a classroom setting that would not occur otherwise?

- Would incorporating specific social networking sites into the classroom setting pressure non-users to adopt a practice that they might not otherwise consider?

- Do the third-party or personal-event components of the social networking sites' user interfaces result in more off-task activity when used in a classroom setting than would occur with learning management systems (LMS), such as Blackboard or Moodle, that don't include such components?

- Does the inclusion of social networking sites in the classroom setting encourage increased formation of within-class cliques?

- What are your views as an instructor on using social networking sites in classroom settings?

- What are the implications of the above questions for educational leaders as they craft and write new and innovative programs, including blended and flipped classroom approaches?

DE Stories

Dr. Nicol R. Howard's Perspective

Back in 2012, I decided to use social media in my teacher education courses. As a former K–12 teacher and a current professor, I grappled with the idea of creating a Facebook account to communicate and become "friends" with my students. I was less concerned with using Twitter because of the difference between "following" and "friending." In my youth I was told by parents and teachers, "I'm not your friend." They said this to distinguish between the roles of friends and adults in the rearing of a child. An assumed level of respect was maintained between teacher-student and parent-teacher when such boundaries were made clear. Today's shift in learning environments to learner-centered classrooms thus raises two questions:

- Do educators now want to be friends with their students?
- Do students actually prefer not to be friends with their teachers?

Although some educators are reluctant to "friend" students, students exhibit a similar response to "friending" faculty (Karl & Peluchette, 2011). College students regularly posting to their Facebook pages might suggest that they would find it convenient to use the same social networks for communication in courses. A study examining faculty and student use of social networks found that 47% of undergraduates surveyed thought it was convenient to use Facebook in the classroom. Although 27% of everyone surveyed welcomed the opportunity to connect with faculty/students online, 22.5% considered Facebook personal, not for education, and 15% articulated apprehension about the potential violation of their privacy (Roblyer, McDaniel, Webb, Herman, & Witty, 2010).

Despite concerns about using Facebook in the classroom, I created what used to be called a group for my students to join in one course and implemented the use of Twitter in another. Facebook use was optional; the use of Twitter was required. Although students were strongly encouraged to create their own Twitter profiles, I did not require that they follow me. Instead, a consistent hashtag at the end of each Tweet established a timeline and record of weekly discussions. Students responded to questions and their peers' responses, all in a succinct 280 characters or less. They received instructor responses only when their Tweets specifically addressed the instructor—an effort not to interfere with their choice to communicate directly with a peer instead of faculty.

Students seemed to prefer Twitter as the social network used in the classroom; however, it is important to consider that this could result from students not being required to friend or follow faculty, as well as the fact that Twitter was required and Facebook was not. That said, the removal of the pressure to friend or follow was an important consideration when using social networks in a teacher education course. Hashtags offered a good way to organize, follow, and record class discussions conducted on Twitter without forcing students to follow the instructor or their fellow students.

Although I have personally moved away from using social media in my current courses (due to privacy concerns), trends in educational technology suggest that the list of social networks available for use in classrooms will continue to increase. I anticipate the continued introduction of new tools worth using with university students in the near future (e.g. Flipgrid), but I encourage an assessment of student preferences, your pedagogy priorities, and an exploration of the tool's privacy settings and efficiency prior to adoption.

Personal Broadcasting

As a society in the early 2000s, we had just begun to see an increase in user-created content with the rise of Web 2.0 tools, which placed "the power of media creation and distribution firmly into the hands of" learners (NMC, p. 9). These tools allowed users to create, comment on, and share content in ways never before experienced and gave rise to the popularity of **personal broadcasting**. According to the Horizon Report, personal broadcasting had "roots in text-based media (personal websites and blogs), [and] is expanding to include audio and video, as the tools for capturing and sharing those media become smaller and better" (2006, p. 11). Podcasts, vlogging, and webcasts were identified as practices that were gaining steam in the higher education community. Similarly, in 2016 Hughes et al. noted an increase in the use, creation, or modification of digital video. In describing grassroots video, the Horizon Report authors heavily emphasized equity:

> From user-created clips and machinima to creative mashups to excerpts from news or television shows, video has become a popular medium for personal communication. Editing and distribution can be done easily with affordable tools, lowering the barriers for production. Ubiquitous video capture capabilities have literally put the ability to record events in the hands of almost everyone. Once the exclusive province of highly-trained

professionals, video content production has gone grassroots. (2006, p. 10)

Students can now easily broadcast their own grassroots video through a wealth of channels—from the venerable YouTube and Twitter to Facebook Live, Snapchat, and more. No matter which venue students choose, the authors of the Horizon Report pointed out the power of having access to an authentic audience:

> Sites that allow easy upload of images, video, audio, and other media also provide students with a low-cost, low-risk means to publish their work as they develop their skills. Comparing their own work to that of others can give students a valuable perspective on their own abilities and inspire them to try new ideas or techniques. (2007, p. 9–10)

The need for students to be content creators and not just passive consumers of education was a concern raised years ago and one that still exists today. Interestingly, preservice teachers are expected to create and curate lessons in their preparation programs. Adding the expectation that they explore personal broadcasting primes them for preparing their future K–12 students to do the same.

What Would You Do?

With a wide array of social media tools and professional learning communities at their fingertips, teachers are using these spaces to share information and resources and to engage in online professional development. Pause for a moment and consider the political climate of today. Now, think about the role social media has played in the dissemination of information.

In a "follow me" and "friend me" culture, *What would you do?* How would you (or wouldn't you) facilitate instruction that supports your preservice teachers' acquisition of knowledge in their professional use of social media for teaching and learning? What lessons would you need to teach in order for them to be prepared for the possibility of teaching with social media in K–12 classrooms?

You might consider one or more of the following thoughts:

- Discuss benefits of building an individual personal/professional learning network (PLN).

- Model Twitter chats and encourage new Twitter users to start lurking.

- Encourage your students to reach out to people they connect with, yet tell them to be critical consumers of content posted by people posing as EduRockStar and remain leery of EduRockStar lists (especially those lacking diversity).

- Promote and model representation of all kinds, and explicitly discuss the difference between representation and tokenism.

- Discuss the importance of multiple viewpoints as opposed to an echo chamber and groupthink.

Join the conversation on Twitter using #DigEquityBook, and tell us what you would do.

DE Wisdom

Bringing SNSs [social networking sites] into the classroom has the potential to enhance some of the more desired elements of college classroom experiences, such as collaboration, motivation, networking, technology savvy, and expansion of content discussion beyond the classroom walls. At the same time, social networking sites have the potential to bring unwanted elements to the classroom, such as antisocial behavior, unneeded distractions, and inequitable educational opportunities based on socioeconomic and cultural differences.

—*Keith Howard and Nicol Howard, 2014*

Considering the possibilities for collaboration, motivation, and networking, we encourage you to leverage the power of social media. For example, invite guests into your classroom via a course-specific hashtag using your alpha-numeric course number (e.g., #edci530, #EDSE457, #MALT603, #IES205). Highlight virtual events like EdChange Global (#ECGlobal) and #EdCampVoice. A key takeaway when you leverage social media is the increased opportunities for you and your preservice teachers to collaborate with other educators, locally and globally.

Remember, too, that it remains important to choose wisely! Not only should you choose the tool wisely, but the activity you select should also be chosen carefully to ensure all in your class are comfortably engaged when using social media. To avoid the fear factor or regression to a quiet face-to-face classroom, build in collaborative conversations and debriefs to re-engage those who may be less social. Also, don't forget that you may have preservice teachers who live in rural areas so it is perfectly acceptable to bring the hashtag to the campus. For example, make the first 30 minutes of class a hashtag discussion and give your preservice teachers the option to participate from anywhere on the university campus. This allows those without devices the opportunity to participate from a campus computer lab; those with mobile devices and poor Wi-Fi connectivity at home can utilize the campus Wi-Fi. Leveling the playing field in this regard is an act of digital equity.

If you have tips to share on this topic, please share them on Twitter using #DigEquityBook.

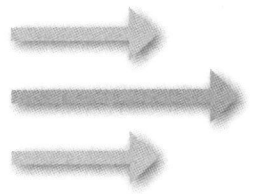

Chapter 3

EXPLORING DIGITAL LEARNING TRENDS

Preservice teachers, as novices, participate peripherally to learning how to teach by observing the expert, their instructor, which provides one model for teaching that influences their developing knowledge, skills, and attitudes of teaching ... Thus, teacher educator choices in using or not using instructional technologies creates a set of digital technology experiences that embody the new literacies ... influencing preservice teachers' decision-making in PK–12 classrooms as professionals.

—Hughes et al., 2016

As noted in Chapter 2, the New Media Consortium (NMC) stimulates and furthers the exploration and use of new media and technologies for learning and creative expression. Examining technology changes over time helps with understanding the need for digital equity practices throughout all sectors of education, because the blind adoption of an emerging technology sets the tone for an inevitable unbalanced teaching and learning experience for all.

The adoption of new technologies calls for a shift in how educator preparation programs support teachers for K–12 classrooms. While Chapter 2 focused on creative expression with social media and personal broadcasting, in this chapter we will look closely at the following learning-based trends:

- **Online Learning:** Learning management systems, flipped learning, adaptive learning, and open educational resources
- **Motivational Innovations:** Gaming, augmented reality, virtual reality, and makerspaces

Finally, we'll take a step back and reflect on the implications of the changes in technology since the NMC's first Horizon Report in 2004—and what they mean as we move forward.

Online Learning

Online learning, first appearing in the Horizon Report way back in 2005, emerged as a way to allow students to receive instruction online due to such factors such as homeschooling, medical conditions, credit recovery, or even extended learning (or professional learning) opportunities. Today, both K–12 students and preservice teachers may actually engage in online learning opportunities to take advantage of the flexibility to learn at any time, anyplace, and at their own pace. Typically, online learning requires a platform, such as a learning management system (LMS), analytics for assessment, and educational content to operate fully.

Learning Management Systems

According to Hughes et al. (2016), **learning management systems** (LMS) have been used in higher education at least since

2004 and first appeared as a trend in K–12 in 2010. Researchers found that 59.4% of preservice students at one university reported their instructor's use of the Blackboard learning management system as a fairly common use of the technology. One student said:

> In one of my classes the assignment was to read an article, write a short essay on a question that was discussed during the class, and post it on the discussion board. Then read and comment on two different essays from our peers by the end of the week. (Hughes et al., 2016, p. 196)

Likewise, in 2012, the K–12 Horizon Report discussed Google Apps (at the time of this writing known as G Suite) and how they integrate into learning management systems. Although LMS had been used for many years in higher education, flexible learning spaces that support third-party integrations and meet universal design standards were beginning to play a larger role, according to the NMC. These new systems would further integrate with third-party tools that allowed for flipped learning, adaptive learning, and open content, as well as gamification (more on this in the "Motivational Innovations" section).

Flipped Learning

An implementation of online learning, **flipped learning** is a strategy that allows learners to complete work outside of class time through "watching video lectures, listening to podcasts, perusing enhanced ebook content, and collaborating with peers" (NMC, 2016, p. 36). Ideally, during class time, students work on challenges and real-world applications to learn authentically, with more personalized support from the instructor. Introducing flipped learning into teacher preparation courses may allow for more hands-on learning and collaborative projects during class time. For example, a professor may assign preservice teachers a Ted Talk related to a specific instructional strategy for viewing outside of class so that more time can be spent in class demonstrating how to apply the strategy.

 Reflect & Consider

Flipped Learning

In an article originally appearing on Easybib.com, Thomas (2015) urged educators to ask five questions before deciding to implement flipped learning, to avoid creating a larger gap in digital equity. Although viewing the videos on a dedicated device in a location with high-speed, consistent internet access is ideal, it is not the reality for all students. The following questions will help you think about alternatives for students in rural areas or locations with poor connectivity:

- *Do your preservice teachers have access to smartphones?* (Also consider whether that access might be to a shared rather than dedicated device.)

- *Do your preservice teachers have access to public Wi-Fi?* (For example, could they access a hotspot in an apartment complex or library?)

- *What other tools do your preservice teachers have available at home?* (Educators can burn DVDs containing the content if students have players at home.)

- *Does your campus have high-speed internet access?* (Students could use campus internet access before, after, or during class.)

- *What resources does your campus have available?* (Schools and universities can invest in portable hotspots and devices for checkout from the library.)

Adaptive Learning and Learning Analytics

Educators can use online learning as more than a source of assignments. To this end, the NMC highlighted **adaptive learning technologies**, "software and online platforms that adjust to individual students' needs as they learn" (p. 44) in the 2015 Horizon Report. These technologies were apparently a continuation of learning analytics, "the interpretation of a wide range of data produced by and gathered on behalf of students in order to assess academic progress, predict future performance, and spot potential issues" (NMC, 2011, p. 28). EDUCAUSE addressed the importance of **analytics technologies** stating that, "understanding how to use new data tools and enabling analytic skills, including data literacy, computational thinking, and coding, are essential for both faculty and students to advance the understanding and use of big data" (2018, p. 8). Preparing preservice teachers to use these technologies in their future classrooms is essential (Figure 3.1).

Figure 3.1 Preservice teachers can collaborate using new technologies, such as Little Bits, with coding platforms.

Open Educational Resources

Open content, currently more commonly known as **open educational resources** (OER), made its first appearance in the Horizon Report in 2010. OER are open licensed, publicly accessible materials and resources that are free for all to use. This practice had huge implications for equity:

> Part of the appeal of open content is that it is also a response to both the rising costs of traditionally published resources and the lack of educational resources in some regions, and a cost-effective alternative to textbooks and other materials. As customizable educational content is made increasingly available for free over the Internet, students are learning not only the material, but also skills related to finding, evaluating, interpreting, and repurposing the resources they are studying in partnership with their teachers. (NMC, 2010, p. 13)

Thus, not only did OER provide financial benefits to schools and students, they also allowed for the remixing of content, collaboration, and professors learning alongside students. The use of OER in educator preparation programs is most visible through a faculty member's selection of course materials. Although this approach offers a no-cost alternative to requiring textbook purchases, it also demonstrates a digital equity strategy preservice teachers may enact when seeking to bring high-quality supplemental learning materials into their future classes.

In addition to OER, the NMC Higher Education Report mentioned massive open online courses (MOOCs) in 2014. MOOCs were created with the vision of global equity in mind, as they were "designed to provide high quality, online learning at scale to people regardless of their location or educational background" (p. 11) as long as one as one had access to the internet. MOOCs were designed to be free. Although often confused with

OER, MOOCs do carry an optional fee, if the learner wants university credit. MOOCs are not necessarily used by faculty in courses; however, university programs may consider MOOCs as a cost-effective alternative for formal and informal teacher preparation.

Motivational Innovations

Technology plays an important role in motivating and engaging learners. Over the years, K–12 has taken the lead when introducing new technologies, which presents a challenge for university faculty when attempting to prepare preservice teachers for their future districts. As we continue to explore the trends that emerged from the Horizon Report, it is imperative for key stakeholders in educator preparation programs to take a closer look at technologies more prevalent in K–12 in order to better prepare preservice teachers for the field of education.

Educational Gaming

Game-based learning first appeared on the 2005 Horizon Report, then returned for the first of several reappearances in 2011. According to the NMC, "the three most recent cohorts of children—those born in the early 1980s, the early 1990s, and the early 2000s—have grown up in a world where digital games have always been an important part of their lives, and entered or graduated from higher education institutions with hundreds of hours of gaming experience" (2012, p. 18). Game-based learning has also been called **gamification**, "the notion that game mechanics can be applied to all manner of productive activities" (2013, p. 20) for engagement and reinforcement of concepts. Under this category, the Horizon Report mentioned **badging**, "a system of recognition that allows students to accumulate documentation of their skills" (2013, p. 22), very similar to merit badges earned by scouts. The assumption embedded in the definitions of gamification and badging is not new, but *is* powerful: Using the same tools in teaching that are used in people's everyday lives is motivational to learning. As the NMC explained:

> Proponents of game-based learning in higher education point to its role in supporting collaboration, problem-solving, and communication, the 21st century competencies needed by American students outlined by Secretary of Education Arne Duncan in late 2010 in the National Education Technology Plan. Advocates also underscore the productive role of play, which allows for experimentation, the exploration of identities, and even failure. Gaming also contributes to the development of a particular disposition well-suited to an information-based culture and rapid change. (2011, p. 20)

The 2012 Horizon Report cited the National Education Technology Plan (NETP) as recommending learning games as a form of assessment and also recommended consideration of simulations in education. Teacher educators would be remiss not to consider these recommendations. Incorporating gaming into teacher preparation courses allows teacher educators an opportunity to model the effective use of gaming in education through the use of what has been historically considered a motivational tool for learning. If gamification is not modeled in educator preparation programs, preservice teachers may miss out on learning how to effectively integrate this innovation into their future classes.

Augmented Reality and Virtual Reality

The interrelated technologies of **augmented reality** (AR) and **virtual reality** (VR) are two more examples of entertainment from our everyday lives being put to use in education. Their first appearances in the Horizon Report were in 2010 and 2007, respectively. The NMC explained fundamental differences between the two:

> AR aims to blend reality with the virtual environment, allowing users to interact with both physical and digital objects. VR enables users to step into an immersive, computer-simulated alternate world where sensory experiences can occur. Head-mounted devices such as Oculus Rift can deliver both AR and VR experiences. (2015, p. 40)

In other words, augmented reality allows the user to interact with simulated projections onto the physical world through a device, while virtual reality re-creates a 360-degree view. Virtual reality is commonly used when cost or travel is a factor. For example, faculty can take preservice teachers on a virtual fieldtrip when a site is prohibitively expensive (such as a foreign landmark) or physically impossible (such as outer space or the ocean floor) to visit. In turn, preservice teachers may consider a similar immersive experience for their future students, especially in a district with limited funds for physical fieldtrips or a remote location. These tools have educational and entertaining implications for consumers and students alike, often immersing the user into learning in unique ways (Figure 3.2).

Mixed reality, "where digital and physical objects coexist also appears as a trend. This hybrid space integrates virtual technologies into the real world so that viewers cannot distinguish where one world begins and the other ends" (EDUCAUSE, 2018, p. 10). What's the difference between virtual reality, augmented reality, and mixed reality? Although they do have many similarities, they all have unique differentiating features. Virtual reality creates a new world in 360 degrees for the user. Augmented reality adds a layer to the physical world, and mixed reality adds an element of interactivity to augmented reality. Aside from simply introducing AR, VR, and mixed reality to preservice

CHAPTER 3 • Exploring Digital Learning Trends

teachers, how can all of this fit into teacher preparation? Researchers at New York's University at Buffalo have worked with a public charter school to create K–12 classroom scenarios, using virtual reality, meant to help preservice teachers navigate various classroom management situations (Loewus, 2017). With more free and affordable tools on the rise, preservice teachers should also be allowed opportunities to model ways to bring AR, VR, and mixed reality into their own instructional design.

Figure 3.2 Preservice teachers at the University of Redlands immersed in learning using Merge Cube VR to understand the experience of their future K–12 students.

Makerspaces

Makerspaces, "also referred to as hackerspaces, hack labs, or fab labs ... community-oriented workshops where tech enthusiasts meet regularly to share and explore electronic hardware, manufacturing tools, and programming techniques and tricks" (p. 40), were first featured in the 2015 K–12 Edition of the Horizon Report. According to the NMC:

Makerspaces are categorized as a development in technology because they are enabled by tools such as 3D printers, laser cutters, and animation software; however, the real focus transcends technology, emphasizing the deep learning experiences and outcomes generated through hands-on activities. (p. 40)

Furthermore, the NMC listed the following as additional staples for makerspaces: Raspberry Pis, Arduinos, Makey Makeys, Adobe Creative Suite software, and sewing machines. **Robotics** are also commonly incorporated into makerspaces to engage K–12 students in hands-on learning that promotes critical and computational thinking. Robotics were projected for widespread adoption by 2021 (NMC, 2016). Although adoption of robotics is not as widespread as other technologies, it is never too early to immerse preservice teachers in learning opportunities that expose them to these technologies (Figure 3.3).

Figure 3.3 Preservice teachers at Chapman University exploring Lego WeDo as an entry into teaching robotics.

Educators are increasingly using makerspaces as a method for fostering creativity and higher-order problem solving, and learners are addressing real-world challenges with innovative solutions (NMC, 2015). Makerspaces continue to spread virally worldwide. The NMC stated that public and school libraries were global leaders in this initiative. A natural fit for makerspaces in educator preparation programs may be in mathematics and science methods courses, yet we encourage the inclusion of maker skills in other courses as well, due to the potential benefits of learning how to promote critical thinking skills across disciplines.

 DE Stories

Dr. Stephanie Quan-Lorey's Perspective

In an effort to help improve the technology integration and preparation of teacher candidates, two of my colleagues and I embarked on a grant to help us identify the gaps of technology usage in teacher education. A survey was shared with teacher education faculty, preservice teachers, and in-service teachers in which the results revealed the need for support and instruction in faculty-driven technology tools and student-centered technology tools. As part of our grant, we decided to pilot some of these technologies in one of the multiple-subject teacher education courses being offered in our department.

Although the course for which I was going to present a classroom technology tool was focused on teaching social studies, it was important for me to introduce a tool that would allow the teacher candidates to bridge their social studies instruction to the STEM subjects (and across most, if not all, of the other curriculum). This led me to select

Scratch (scratch.mit.edu), a programming language created by the Lifelong Kindergarten Group at the MIT Media Lab, as the tool I decided to preview.

As the interests and skills of coding have become more prominent in society, many students' desires to learn how to program have piqued. As a result, several teachers have explored methods to integrate this into their instruction in a user-friendly way both the students *and* teacher can understand. Priding itself on its capabilities to be utilized by a learner of any age, Scratch is centered around the idea of creating scripts by stacking blocks together. Each of these blocks serves as a command to carry out animations. The projects one builds on Scratch can be produced for any subject area in education and its limitations of what can be created are dictated only by the user's imagination.

During the class session where I piloted Scratch with the teacher candidates, introductory skills were covered to orient them to the program. The teacher candidates were shown briefly how a Scratch user could create his or her own graphic stories, games, music, art, or animation to design a project, review tool, or any other type of educational product. Key terms, functions, and how to build scripts were explained, and the teacher candidates were given the opportunity to follow along on their own devices as we explored the various facets of Scratch. At the end of the guided instruction, the teacher candidates were invited to work with small groups to create their own Scratch projects or navigate through the projects uploaded by other users in the Scratch community to see some examples of finished products while using a Makey Makey.

Upon my introducing Scratch to the teacher candidates and mentioning the use of Makey Makeys, an initial feeling of nervousness and uneasiness enveloped the room. Several of them assumed coding was going to be difficult and the process of learning how to program using Scratch would be tedious, especially for those who felt their technology skills were low. Through use of a short video to hook them into seeing the possibilities of projects that could be created through Scratch, well-paced guided instruction and practice of some introductory features of building a program, and time to interact with several projects through the use of a Makey Makey, the fear many of them first experienced gradually dissolved as their engagement and creativity heightened (Figure 3.4). The teacher candidates acknowledged the ease and simplicity of introducing coding to their future students using Scratch and were overflowing with ideas of how to integrate its capabilities into their teaching across multiple subject areas.

Teachers sometimes feel uncertain or display pushback for integrating technology into their classrooms because of not feeling confident or prepared to weave it in. I remind the teachers I work with that technology does not have to take over one's instruction, but rather can be utilized to support and enhance teaching. Just as we encourage our students to take risks and accept challenges in their growth as learners by utilizing prior knowledge and experience to help them navigate through their difficulties, this can also apply to teachers in expanding their repertoire of teaching tools by including technology. Use your foundational knowledge of excellent pedagogy and weave it into your integration of technology in the classroom. Incorporating technology does not strip away these core ideas of good learning and teaching and should not leave the teacher feeling paralyzed or scared to use it. Let it serve as a way to cultivate a student-centered learning environment where engagement and collaboration meet the digital universe.

Figure 3.4 Despite initial nervousness, preservice teachers find collaborating to be a useful strategy when learning new technologies, such as Makey Makey.

What Changed?

In examining the Horizon Report's forecasts, projections, and trends for higher education since 2004, we saw two over-arching themes emerge: the interconnection between the three areas of focus (higher education, K–12, and the outside world) and an innovation paradigm shift away from new tools towards how tools were used.

Interconnections of Education Spheres

Technological adoptions in any domain soon begin to creep over into the others. We've all seen this happen many times throughout the course of the last fifteen years. Although the outside world tends to outpace and influence both higher education and K–12, the dynamic between the latter two fields is an interesting one to watch. Up until about 2014, higher education appeared to be on the cutting edge, with K–12 education following soon behind; however, K–12 schools were first to implement BYOD on a large scale. K–12 also led the way with makerspace adoption.

Although the outside world has always had a lead on educational spaces with technology adoption, we noticed an increase in the speed of adoption in schools and higher educational institutions. For example, although blogging was a top trend of 2004, it was not addressed in the Horizon Report until three years later. Grassroots video followed the same trajectory; for example, innovations, such as the inception of YouTube in 2005, were not reflected in the report until years later (2008 for YouTube's mention). Fast-forwarding a few years, however, shows that wearable technology, a mainstream trend in 2013, was referenced in the Horizon Report the following year. This can possibly be attributed to the spread of information through social networking, which helps to communicate best practices at a rapid pace.

New Uses versus New Tools

Perhaps most significantly, the second trend was an implicit paradigm shift surrounding innovation. In the earlier years of the report, as well as in society at large, each year seemed to bring something new. Contrary to our expectations, we rarely found that innovations progressed in order of predicted use from long-term, to mid-range, to short-term. Rather, the list seemed almost random, with certain technologies disappearing and reappearing a few years later. Eventually, however, we began to see certain items recur on the list. Many of these themes included devices (whether mobile or tablet), social networking and collaboration, and learning with web-based technologies, or any combination of these elements.

In our review, 2008 was the first year that we noted a decline in the attention paid to new gadgets and platforms, with more press highlighting new ways that people were using recent innovations. Before 2008, we witnessed the launch of smartphones and other mobile devices, as well as social media giants Facebook, YouTube, Twitter, and many more, which would change how we interacted with the world. Afterwards, new platforms and inventions would leave small ripples, but few would make as deep a splash as those prior to 2008. However, we did notice a shift in how people used these technologies. For example, who would have known that a platform created for the social lives of college students could one day potentially skew a presidential election? Who could have guessed that a 140-character (now 280-character) snapshot into "what's happening" would one day tell the story of a city aching for justice after yet another tragedy? All of this to say, no one truly knows what's on the horizon.

What Would You Do?

Imagine learning about a new app or technology tool that was recently adopted by several local districts. You've witnessed a positive reception to the adoption by the teachers at one K–12 campus and less enthusiasm at another. When you inquired about the adoption with the administrators and teachers at both sites, you learned that the enthusiastic teachers were offered on-going professional development and those with less positive attitudes were not. With all of this in mind, *What would you do?* Would you still consider introducing this tool in your educator preparation course or professional development session? What would you consider in advance?

So, we all know how rapidly new technologies emerge. Sometimes this thought alone is off-putting and discourages teachers from learning a new tool only to change to another in short order. In a scenario such as this, you might introduce the tool early in the semester and offer multiple opportunities for your preservice teachers to utilize the new technology. You might even consider scaffolding and differentiating through collaborative grouping strategies and/or flipped lessons. Most importantly, at the conclusion of your activity using the new technology you might consider a guided discussion with your preservice teachers about what you witnessed on the K–12 campuses to ensure they make wise decisions about the technologies they adopt in the future.

Tell us what you would do on Twitter using #DigEquityBook.

DE Wisdom

> I wonder when the shift from seeking resources to assist in teaching in learning, became "there is one company that will be *the* tool/resource." My concern is that when we focus on the resource itself, we lose sight of the reason we sought it out in the first place: to assist in facilitating learning.
>
> —*Regina Schaffer*

Enacting digital equity calls for placing learners, in this case preservice teachers, at the center of the learning experience by focusing on what they may need to be prepared for their own future classroom. This serves to model best practice, while engaging the preservice teachers in authentic and relevant learning of their own. The technology serves as a tool to enhance the learning experience, not the focus, which is a strategy that will hopefully remain at the core of their decision-making related to teaching and learning.

If you have tips on how to focus less on the technology and more on achieving digital equity, please share them on Twitter using #DigEquityBook.

Chapter 4

OVERCOMING THE BARRIERS

> *In recent years, educators have witnessed an unprecedented acceleration of new and innovative technologies. It is not uncommon for educators to have differing opinions about which tools are helpful in their classrooms and which may bring unnecessary complications.*
>
> —Howard & Howard, 2014

Overcoming the barriers in pursuit of digital equity does not occur overnight, nor does it happen in isolation. There is no shame in acknowledging the challenges, or even collaborating with faculty and colleagues, to tackle the ones that seem to be the most insurmountable. In this chapter, we will outline several of the challenges and share stories from the field that offer practical examples of how to best support preservice and in-service teachers in their acquisition of new technology knowledge and their pursuit of digital equity for their future and current K–12 students.

CHAPTER 4 • Overcoming the Barriers

Diversity and Digital Equity

What are the connections between diversity and technology when seeking digital equity? According to Rafranz Davis, "Diversity matters in EdTech because not all tools, devices, apps, and ideas are created equally—nor are the learners" (2015, p. 5).

Inevitably there comes a moment when teacher educators must unpack their own biases and should continue to do so when evaluating digital tools for teaching and learning. The most obvious quality to look for is whether the tool is user-friendly and the less obvious includes issues related to diversity. As educators explore ways to enhance the learning potential for all students, it is imperative to remember that certain populations of students have different needs. These needs will have to be addressed if educators are to increase the possibility of students benefiting from technologies that will impact their learning process (Chen, 2007). Gender and racial bias, as well as subtle and overt visual misrepresentations and stereotypes, contribute to the diversity and technology issue we will address in this chapter.

 An Open Letter
from Technology Specialist Patricia Brown

Dear Teacher Educator,

Is diversity just a buzzword?

Diversity has become a buzzword and a commonly discussed ed tech topic in Twitter chats, educational conferences, and PD sessions around the world. At this point, we should be going beyond surface-level diversity chats, and instead have deeper conversations on how we can create diverse *and* inclusive educational spaces where educators are equitable in their mindset, and practice, and are willing to gain the skills they need to embody and live out the idea that all kids can learn. These conversations are important because in this world there are too many ways that bias creeps into our lives: the media, our community, our families, our friends, and our experiences.

How do we move beyond these biases?

How do we identify our own biases and address them in a healthy way?

It starts with you as an individual with an understanding that you need to decenter yourself—ego aside ... and reflect. There are many resources available to begin this very personal work, but we all can begin very simply by reflecting on these questions:

- How does my own social location (race, class, gender, religion) shape my mindset about teaching and learning, the students I am serving, and the practices I act out?

- What more do I need to learn about the things I don't know related to culture, power, and difference?

- Where can I learn that material?

- How can I be a more critically conscious educator?

Creating diverse and inclusive spaces allow for people to gain so much. Learning from each other is what makes us grow. How do you look internally and unpack your own implicit or overt biases?

Gender Differences and Racial Bias

Since the 1980s, studies have shown that the advertising images of computers have been predominately male-oriented (Cox, 2009), and people of color have been less represented or depicted in less empowered roles. Concerns about gender stereotypes in children's literature have also been reported, dating back to the 1960s. Motivated by an interest in examining whether these same stereotypes appeared in digital imagery, Bradshaw, Clegg, and Trayburn (1995) investigated gender bias in educational applications used in elementary schools and explored the messages young children were receiving from computer screens. They discovered that similar biases exist in the use of computers, as well as the imagery deployed in the educational software. There was some evidence that when "androgynous" humanoid figures were used, students most often defined the characters as male, further suggesting that gender bias exists in the design of educational software.

Another study by Sheldon (2004) almost a decade later, indicated that significantly more male characters than female characters were seen in educational software, making it difficult for teachers to address gender diversity when programs suggest boys are more valued than girls. Additionally, research in several countries, including the U.S., has shown that girls have less computer experience and have less comfortable attitudes toward computers than boys (Hanson, 1997; Sanders, 2005). Besides girls exhibiting less comfortable attitudes toward computers, a study by Howard, Sauceda Curwen, Howard, and Colón-Muñiz (2015) revealed that girls were significantly more uncomfortable than boys with using social networking sites as a communication tool in school. Exercises related to the critical analysis of new programs and apps need to be integrated into educator preparation (including professional development) in order to raise the level of awareness of gender bias that can potentially influence damaging subtle or overt stereotypes.

Recent data and research suggest that today's students bring culturally and ethnically diverse backgrounds into classrooms. In addition to gender differences and prevalent stereotypes in the design and use of technology, racial bias also exists. Historically people of color have been the victim of negative imagery; the remnants of these constructions continue to have contemporary influences in schooling experiences in the U.S. (Howard, Flennaugh, & Terry, 2012). For example, educational apps have been designed in recent years without all of the faces and hues that represent the students or teachers of color in today's diverse classrooms. Additionally, certain imagery used in online and app simulations depicting the lives and experiences of enslaved people of color often misrepresent their important histories. Within the context of digital equity, racial bias can be seen in the representation, lack of representation, or misrepresentation of students of color in educational software and apps.

Addressing the challenges that exist when educating our increasingly diverse learner population is a priority in K–12 and educator preparation. Likewise, an awareness of cultural and ethnic backgrounds of both K–12 students and preservice teachers is an important step in this process. In response to the obvious concerns related to gender and racial biases and the impact on digital equity, as well as the fact that students are increasingly exposed to educational technology in schools, the International Society for Technology in Education drafted an updated version of the ISTE Standards for Educators in 2017. These standards should be used not only by current K–12 teachers, but also by educator preparation faculty and instructors if for no other reason than to model digital equity practices. Navigating the challenge of minimizing gender and racial bias begins in educator preparation not only with pedagogy, but also with a change in the selection of digital content and applications.

Assumptions About Teacher Tech Knowledge

Dynamics in K–12 classrooms have changed radically over the years with students from diverse backgrounds and varying levels of knowledge about technology. Educator preparation must change as well in order to better support preservice teachers in their development of the necessary skills required to meet the needs of students today. There is a great deal of enthusiasm about the adoption of new and innovative technologies to aid in teaching and learning, hence educator preparation programs would be remiss not to include the same technologies used in K–12 classrooms. Likewise, faculty in such programs must be sensitive to and avoid assumptions about the technology skill and knowledge levels of preservice teachers (just as teachers must avoid assumptions about students in their K–12 classrooms). To adequately prepare future teachers and disseminate new information to current teachers, we must be aware of their true technology knowledge-base. Furthermore, it is imperative to remember that using technology efficiently for personal use does not equate to an expert ability to teach K–12 students how (and when) to use technology.

 Reflect & Consider

Avoiding Making Assumptions

Simply because a preservice teacher is a millennial or from the iGen, their professors should not assume that the student possesses a vast amount of knowledge about how to use technology for teaching and learning. How will you avoid making these assumptions? Here are a few thoughts to consider:

- Anonymously survey your class on day one using an app, such as Poll Everywhere, or Kahoot!, to determine your students' level of technology knowledge. Consider asking questions on scales (from 1 to 5). For example, you could ask, "On a scale of 1 to 5 (with 5 being most), how comfortable are you with creating a Google Slides presentation?"

- Differentiate the questions you ask to give preservice teachers a chance to indicate how well they know a particular tool or their confidence levels in learning a new tool. You may ask a question similar to the one in the previous bullet to determine how well they know a tool. A good question to ask related to confidence level may be, "On a scale of 1 to 5, how comfortable are you with learning new web-based tools?"

- In addition to asking questions on the first day, consider periodic check-points where you ask for feedback on your technology-rich lessons to ensure you aren't leaving anyone behind.

- Observe your preservice teachers while working on projects to determine how well they know certain tools and how comfortable they are with using these same tools. Survey responses often can be optimistic views of our true capabilities or underestimates of our confidence.

▲ ▲ ▲

Quality Teaching

Quality teaching is imperative today (and forever) as it ensures that all students have access to the same learning opportunities, the same information, and the same chances to create, collaborate, and achieve (Burton, 2016). Learners, even in educator preparation programs, should be offered an environment where digital resources and technologies are integrated into their formal and informal learning experiences beyond direct instruction. An environment such as this can open the door for innovation and lead future teachers to a better understanding about how to facilitate similar learning experiences for their own K–12 students. For example, preservice teachers can be offered opportunities to explore new technologies with follow-up conversations about how to integrate those tools into future classrooms (Figure 4.1).

Figure 4.1 Preservice teachers explore Lego Mindstorms for the first time, before discussing how and where to integrate robotics into K–12 instruction.

All of that said, poor teaching is an obvious barrier in pursuit of digital equity. "Tech is an accelerator … If you apply it to bad, ineffective practices, you get faster, bad, ineffective practices. If you apply it to good, high-quality teaching practices, you get faster, higher-quality teaching practices" (Culatta in Molnar, 2014). Teacher education professors should reflect on their own practices to ensure quality teaching, while encouraging preservice teachers to do the same. So, where does quality teaching today start exactly? With all of us; we must strive to:

- **Employ strong pedagogical practices** that support the content knowledge and development of all learners regardless of race, ethnicity, national origin, gender identity, disability, English language ability, religion, socioeconomic status, or geographic location.

- **Recognize the nuanced needs of learners** related to digital citizenship.

- **Ensure sufficient access to devices and the internet** in your classroom, and effectively plan instruction so that all learners have opportunities to innovate using digital resources at key points throughout the academic year.

- **Select user-friendly tools inclusive of diverse figures or icons** that represent the demographics of all learners.

- **Offer personalized learning opportunities** that are inclusive of rich technologies for innovation.

A critical component of teacher preparation is the consistent modeling of quality teaching by faculty. When speaking of quality teaching in the context of digital equity, faculty should also consistently model the use of technology as this influences preservice teachers' perceptions of using technology for their own teaching (Albee, 2003; Darling-Hammond, Meyerson, LaPointe, & Orr, 2009; Hsu, 2012; Mills, 2014). *Why is this important?* According to the ISTE Standards for Educators (Leader, indicator 2b; 2017b), it is essential that educators advocate for equitable access to technology, digital content, and learning opportunities that support the learning of all students.

DE Stories

Dr. Betina Hsieh's Perspective

I've always thought it was essential to support teacher candidates in integrating technology, multi-modal and digital literacies, and critical thinking into their practice, beginning with preservice teacher education. At first, this integration came through a "21st centuries literacy project" in my secondary reading and writing preservice course, in which teacher candidates themselves created digital products (e.g. blogs, digital stories) or engaged in digital interactions (e.g. Twitter chats) relevant to their future professional contexts. This was based on the philosophy that teacher candidates would benefit from engaging with technology in professional ways that could directly apply to their classrooms. Although this experience proved powerful for many of my teacher candidates, many others didn't really seem to see connections between their reflections or professional conversations and their future classrooms.

Because of this, I wanted to more directly connect technology integration with the central focus of the course: lesson planning. To do this, initially, I just required students to embed technology into their lesson plans, however, the technology integration I received ranged from the use of PowerPoint presentations by the teacher to various uses of the Google Suite to students creating digital stories. Given this range, I chose to integrate use of the ISTE Standards for Educators or ISTE Standards for Students as part of the lesson plan requirements. Credential candidates first chose focal ISTE Standards (as well as their content standards and language objectives), then they designed content that integrated technology more intentionally. For each use of technology, candidates were required to relate their pedagogical choices to the ISTE Standards, with the option to integrate a secondary framework, such as the Substitution Augmentation Modification Redefinition (SAMR) Model or P21's Framework for 21st Century Learning, in their rationale.

Since integrating the ISTE Standards for Educators and for Students as part of this lesson planning assignment, candidates have demonstrated a deeper understanding of the uses of technology, both as a tool that can substitute for traditional paper and pencil tasks, *and*, more importantly, as a tool that can transform the learning opportunities and educational spaces of the classroom, for themselves as professionals and for their students. They've done this through the integration of online labs into their curriculum, engaged collaborative opportunities that empower students to create their own digital products, increasingly multimodal instruction and assessments, and general engagement with new tools themselves through a pedagogical lens. The integration of the ISTE Standards to ground these lesson plans has allowed abstract concepts to become real, and it has led to real technology integration and implementation in classrooms. Once in their own classrooms, several of my former students Tweet me to describe the ways that they are using technology powerfully in their instruction, assessment, and communication with students and families.

Reflect & Consider

Key Questions About Quality Teaching

- Do your technology choices support your pedagogical practices and the development of all future (or current) teachers you are preparing for the field?

- Does your use of technology model digital citizenship and support the nuanced needs of your learners' potential (or current) K–12 students?

- Do you ensure sufficient access to devices and the internet in your classroom?

- Do you model the selection of user-friendly tools inclusive of diverse representations, imagery, or icons that represent the demographics of all learners?

- Do you encourage future (or current) teachers to innovate using multiple technologies?

Staying Current, Connected, and Curious

Keeping current and connected is a challenge in educator preparation. The true barriers in doing so are related to the fast-paced evolution of technologies and time management concerns when it comes to planning and facilitating instruction. Connecting with other faculty in local educator preparation classrooms, or even classrooms across time zones, is also a daunting task. It takes time to identify faculty who teach similar courses or content, as well as to determine the appropriate technologies and develop shared tasks for preservice teachers to complete. If, or when, a collaborating faculty and classroom are identified, the following tips may help overcome potential barriers:

- If there are conflicting lengths or durations of your courses, use an asynchronous form of communication, such as Flipgrid, to connect locally or globally.
- When connecting synchronously using such tools as Skype, Zoom, or YouTube Live, be prepared for the unexpected (e.g., poor connectivity). Setting up a free LMS or content management system, such as a Google Classroom or free Canvas account, with a discussion board that includes prompts specific to what you plan to discuss live, is one workaround. Or consider having your preservice teachers write drafts of their posts off-line while waiting for connectivity to be restored.
- In the event that one class has more preservice teachers than another, consider innovative grouping strategies between your collaborating classes when attempting to partner students. For example, a class of twenty-six may assign two students to work with one student from another class with only thirteen students.

More often than not, remaining curious is the less difficult task in pursuit of digital equity in educator preparation programs. According to Howard (2015), to keep your teaching fresh and effective you should learn about colleagues, build community, expect the unexpected, remain current, and, most importantly, stay curious. Curiosity has the power to keep individuals motivated, and to help new discoveries emerge.

Yes, preservice teachers should be encouraged to remain current, connected, and curious. Due to the nature of emerging technologies, this is just as important for professors in educator preparation programs in order for them to effectively guide preservice teachers in their future pursuit of digital equity.

DE Stories

Dr. Marialice Curran's Perspective

Early in my higher education career, I noticed a pattern that connected educators were K–12 teachers, not teacher educators. Understanding that learning is mobile, immediate, social and collaborative in nature, I knew that teacher preparation programs must model this approach in educating teacher candidates for the reality of connected classrooms in K–12 schools.

Connected educators were embracing these changes and incorporating this new reality into the classroom. In the early 2000s they were using digital technologies and social media, such as Twitter, as an asset to help enrich their pedagogy. As Cooke explained, "The power of Twitter is not Twitter itself; it's the connections it facilitates. Those connections can break the sense of professional isolation that many teachers feel within the walls of their own schools while invigorating their lesson plans by exposing them to a daily global exchange" (2012).

Personally, I became connected in 2010, and I noticed that connected educators were redefining the traditional role of the teacher and student in the classroom into a co-learner model where the focus was on learning collaboratively side by side. In stark contrast, I felt like I was on a deserted island on my university campus. I had one *colleague* who was a connected teacher educator and other than that, we felt alienated by the majority of faculty. Innovative programs, such as my iMentor research, went unnoticed and publications for tenure counted only in peer-reviewed journals.

The iMentor (www.irma-international.org/viewtitle/74916) research happened in my elementary methods course during the fall 2011 semester. I reached out to my PLN (https://mbfxc.wordpress.com/2011/07/11/looking-for-virtual-elementary-teachers-to-be-science-social-studies-mentors) and asked for virtual mentors for my teacher candidates, writing:

> I would love to connect my graduate students with virtual mentors on how to teach science and social studies in PK–6 classrooms this fall semester. It will be a great opportunity for teacher candidates to see the power of global collaboration, as well as an opportunity to learn with you and your students. I envision Skype sessions, virtual science experiments, and social studies projects happening together! If interested, please let me know how I can best contact you. I see higher education changing. Please consider being a part of this exciting virtual mentoring program!

The results (https://mbfxc.wordpress.com/2011/12/15/thank-you-pln-for-changing-teacher-preparation) were fantastic! Teachers around the world volunteered and each week we had a connected educator Skype in to share with our class. Plus, each teacher candidate had to schedule a time for virtual field experience with his or her iMentor. Making these connections was a game changer for teacher education because there was never a guarantee that these teacher candidates would have experienced what a connected classroom looks like during their local field experiences.

That same semester, I had blogged (https://mbfxc.wordpress.com/2011/05/13/high-school-skype-and-twitter-project-request) looking for volunteers to collaborate on my first-year seminar course for incoming college freshmen in Connecticut called, Pleased to Tweet You: Are You a Socially Responsible Digital Citizen? This course which we referred to as #FYS11, taught me more than I could have ever imagined. The course description included the following:

> Schools across the country promise to provide a safe environment for learning, but so many students are afraid and embarrassed to come to school. In today's globally diverse and digital world, a bully's reach goes far beyond the playground. As more teens use computers, cell phones and other electronic devices they will experience being harassed, threatened and humiliated publicly online at greater rates. Cyberbullying is the biggest hazard our young people face today and will continue to face in the future as more teens consume and produce digital media. An interactive multimedia approach to this course will provide students an opportunity to explore the problem and extent of cyberbullying through readings both on and offline. Using a reflective lens, students will create an action plan to help others navigate the Internet as responsible digital citizens.

The collaboration between college freshmen in West Hartford, Connecticut and high-school juniors in Birmingham, Alabama began as an idea for students to have the opportunity, despite geographical limitations, to connect, communicate, think critically, and act creatively via social media to address issues around cyberbullying. As the project evolved, its scope expanded to encompass a holistic approach to digital citizenship in local, global, and digital communities that developed into the iCitizen Project (https://www.mbfxc.com/the-icitizen-project.html).

The iCitizen Project became more than just a school project; it was a transformative experience for both the college freshmen and high-school juniors and developed into an opportunity to change minds, attitudes, and hearts. Through Skype and Twitter, the students collectively defined an *iCitizen* as an individual who is aware, empathetic, socially responsible, and someone who believes in social justice and models being a citizen of the world both on- and offline.

The final multimedia project promoted consciousness and empathy in a digital world. Both classes bridged the gap between two schools separated by geography and created an online iCitizen community. The participants concluded:

> We initially put a lot of emphasis and focus on the issues of bullying and cyberbullying, and while it does remain a large problem both on- and offline, we felt that teaching empathy first is more effective than trying to stop bullying later. Together we learned what it means to be an active citizen instead of just a resident, an enabler of change, and not just a bystander. We learned to humanize the person next to us, around the world, and across the screen. For a generation who has grown up around computers, it's hard to think there's anything new that you could possibly learn about the Internet. But this project has shown us that there is always room to grow, connections to forge, and communities to contribute to, both in your backyard and behind a computer screen. And the tools we've acquired by working together on this project will be used to benefit and educate others to create a much more rewarding online experience for everyone. (Curran, 2012)

CHAPTER 4 • Overcoming the Barriers

What Would You Do?

You are teaching a new course in your program that has never been taught before. You've developed the syllabus and considered both your state credentialing requirements for utilizing technology, as well as the ISTE Standards for Educators. Although you have addressed collaboration through face-to-face activities, you are curious about integrating online collaboration opportunities. *What would you do?* How would you identify other university classrooms to collaborate with on activities?

You might consider reaching out to your professional network to ask for ideas of recommendations of others with whom to connect. If you have a social media presence with a robust educational following, you might consider your personal/professional learning network (PLN) to identify local or international professors willing to collaborate on one activity.

Join the conversation on Twitter using #DigEquityBook, and tell us what you would do.

DE Wisdom

> As you connect (physically and virtually) to your content, colleagues, and curated resources, remember to remain curious and keep it current. We have made some powerful shifts in education over the years, and it is frightening to feel lost or left behind. As a child, my curiosity motivated me to move in a positive direction. Reflect on the times that your curiosity and imagination took you on an amazing journey, and allow that same spirit into your classroom—for you and for your students. When we are curious, we learn and discover new innovations. Allow your curiosity to drive your connections.
>
> —*Nicol Howard, 2015*

We encourage preservice teachers and those who educate them to remain current by reading about innovations from multiple resources including, but not limited to, the books, blogs, and publications of the International Society for Technology in Education (https://www.iste.org). Bring these resources into your courses and allow your preservice teachers to do the same.

Please share what you're currently reading, and your new discoveries, on Twitter using #DigEquityBook.

Chapter 5

INNOVATIVE APPROACHES TO DIGITAL EQUITY

Although the presence of technology does not ensure equity and accessibility in learning, it has the power to lower barriers to both in ways previously impossible.

—U.S. Department of Education, 2017, p. 85

Now that we have examined the changes over time, as well as the challenges and successes related to digital equity, we'd like to look more closely at how to close the digital equity gaps by offering tips for sustainable success. Although the following can be applied throughout educator preparation, we hope the separation of traditional from non-traditional educator preparation tips helps with navigating university department or district policies and meeting institutional expectations.

Traditional Educator Preparation Courses

As Project Tomorrow (2017) found and we stated in Chapter 1, the best way for teacher candidates to understand best practice is to experience it themselves. Riccards (2016) echoes this, calling for personalization: "Prospective teachers are no different than those they will one day teach. So why aren't more teacher education programs focused on the learner, with programs based on the needs and preferences of that adult learner?" Again, personalization can encompass a variety of things, including home culture, as stated in Thomas (2017b). Riccards also called for more blended learning, where preservice teachers would have opportunities to leverage both face-to-face and online learning.

Additionally, teacher education programs need to support lifelong learning. It is not enough to earn a degree and decide to stop improving. As stated in Hodges, Carpenter, and Borthwick:

> If future teachers take a course at the beginning of their program, by the end of the program some of the specific technologies that they would have learned about may no longer be relevant. And certainly by the time they are in the field and have been teaching for two or three years, the technologies available will have changed significantly. The rate of change in technology is so rapid that we have to prepare teachers to continue learning after graduation. (2017, p. 20)

In their journeys to self-guided learning, educators may participate in personal/professional learning networks (PLNs), where they may gain information, feedback, support, and kinship with peers around the world. They will also likely be exposed to a range of information—with varying degrees of accuracy and quality. Therefore, preparation programs should help candidates become critical consumers of information, avoiding the traps of hopping on buzzword bandwagons just because they are popular. Joseph South, Chief Learning Officer at ISTE, elaborated on the need for teacher candidates to know more than what tools are available, but instead what tools are appropriate and useful:

> My sense is that the majority of teacher education programs are more focused on helping teachers to learn how to use technology tools than on helping them make a good selection about which tools to use. It is not that they think that this is not important. It is just not the focus of their courses. (Bull, Spector, Perischitte, & Meiers, 2017)

He continued to advocate for educator preparation programs to include the explicit goal of "developing teachers who are savvy consumers of technology," and suggested collaborative case studies as a possible approach.

Hodges et al. state that, "educator preparation programs must help all candidates understand their essential role as advocates for what is best for students—the role of the citizen scholar" (2017, p. 18). Key components of advocacy include *culturally relevant pedagogy* (CRP) and *culturally sustaining pedagogy* (CSP). The former, a term coined by Gloria Ladson-Billings in 1994, refers to "a pedagogy that empowers students intellectually, socially, emotionally, and politically by using cultural referents to impart knowledge, skills, and attitudes" (in Borrero, Flores, & de la Cruz, 2016, p. 29), which helps to maximize academic achievement (Durden, Dooley, & Truscott, 2016, p. 2). Paris (2012) offered culturally sustaining pedagogy as an alternative term, in an attempt to "support young people in sustaining the cultural and linguistic competence of their communities while simultaneously offering access to dominant cultural competence …

[supporting] multilingualism and multiculturalism in practice and perspective for students and teachers" (p. 95).

In other words, CRP implies that teachers should be representative of all cultures, while in CSP the teacher takes a more active role in bolstering students' cultural identity while navigating the dominant culture. Although this should have always been present, it is now critical for educator preparation programs to prepare candidates with these skills, because:

> Although the demographic makeup of the U.S. student population continues to become more diverse, public school teachers are still predominantly European American. Nationwide, approximately 83% of the teaching force is White, yet over 49% of students are students of Color. Further, in cities like New York, Los Angeles, and Chicago, over 80% of students enrolled in public schools are students of Color. This divide not only calls upon us to change the categories and terms we use to describe "minority" students, but, as educators, this divide also requires us to build our critical consciousness and interrogate the cultural dissonance between teachers and students by naming the racial hierarchies that schools embody and reproduce. (Borrero et al., 2016, p. 28)

However, there is one caveat: When implementing CRP or CSP, it is important to avoid making it "a feel-good or obligatory addition to teacher preparation" (Lynn, 2014) or just treating it as an add-on. Paralleling the practice of representation, it should not be viewed as a box to be checked; instead, it should be purposeful and interwoven throughout the course of the program.

An Open Letter
from Dr. Keith E. Howard

Dear Teacher Educators,

Education technology literature has supported a narrative for many years that the current generation of students is far more advanced in their understanding and use of technology than prior generations, and perhaps even ahead of the teaching workforce (Prensky, 2001, 2011) in their tech savvy. These so-called "digital natives" are presumably better equipped to utilize technological advancements as a result of being born into the digital age and growing up using it in their daily lives. Those who grew up without the technology and had to adapt prior practices to it ("digital immigrants") are presumed to have adapted less to the new technology, leading to different levels of experiences with it in their daily lives.

In more recent years, evidence has emerged suggesting that this technology-use gap is overstated, and that digital-immigrant teachers actually depend as much or more on technology than their digital-native K–12 students (Wang, Hsu, Campbell, Coster, & Longhurst, 2014). A review of 127 articles published on the subject between 1991 and 2014 concluded that although digital natives exhibit high digital confidence and skills, their digital competence may be much lower than their digital-immigrant teachers, and there is a lack of evidence that they even want to use their skills for academic purposes (Gallardo-Echenique, Marqués-Molías, Bullen, & Strijbos, 2015). Moreover, this research suggests that there are several factors more influential than age influencing levels of technology competence.

My experiences in teaching educational technology in higher education, from undergraduate to doctoral levels, over the past ten years have been consistent with the more recent literature cited above. Although my undergraduate classes are noticeably more likely to be filled with students sporting the latest and most sophisticated laptops, tablets, and mobile technologies, their technology savvy when it comes to utilizing these devices often falls short of their confidence and exuberance. Beyond using email, word processing, texting, and social media, few among these digital natives actually separate themselves from the group in terms of productivity or computational thinking. Conversely, doctoral-level classes predominantly comprised of digital immigrants are more likely to be attended by students with work experiences that have required them to learn how to use technology in problem-solving contexts. Whereas they may be less proficient at navigating a social networking app and less familiar with a touchscreen device, they are often more likely to have experiences that have taught them how to utilize a formula in a spreadsheet to solve a problem or how to use a database to organize data and inform decisions.

Access to digital technology is not nearly as important as the purposes for which it is used, or the purposes for which there is the desire to use it. Patterns of unproductive use of technology have been found to produce a "time-wasting gap" for many digital natives (Richtel, 2012), further discounting the potential advantage that mere access to technology might otherwise have afforded them. Efforts to leverage digital natives' interest and proficiency in technologies that they frequently use are intuitively appealing, but have sometimes been met with discomfort on the part of students who prefer not to mix their preferred leisure tools and platforms with their academic lives (Howard, Sauceda Curwen, Howard, & Colón-Muñiz, 2015) and may be accompanied by undesirable risks (Howard, 2013).

For those involved in preparing educators for technology use in the classroom, it is important to convey to pre- and in-service teachers that they often possess more technology competence than they may realize, especially when it comes to utilizing technology to solve problems at home and in the workplace. Digital immigrants' use of technology to solve everyday problems and tasks, such as online banking, preparing tax returns, or planning events is arguably more valuable in preparing students for productive use of educational technology than typing a paper, sending text messages, or posting to social media platforms. As educators, we should embrace the digital confidence of the digital natives for the openness it provides them with when it comes to accepting technology as a part of their learning spaces. At the same time, we should remember that in an educational setting, it is not just what technology tools you are able to use that is important, but rather how you use these tools to enhance knowledge acquisition and capacity to be a productive learner.

Non-Traditional Educator Preparation

We have spoken at length about preservice teachers, those who are in the process of completing their studies prior to being charged with shaping the future. However, what about the rest of us?

Teacher preparation ideally never ends. As Hodges et al. (2017) implied, we should all aspire to be lifelong learners and expose ourselves to multiple opportunities to learn and grow. Some of these opportunities may be facilitated by our districts or institutions, but there are also opportunities that we can (and should) identify independently. In this section, we will examine some areas of particular note, most of which have gained in popularity over the past few years.

Badges and Microcredentialing

If the term *badges* reminds you of scouting merit badges, you're not far off. According to the MacArthur Foundation (n.d.), "digital badges are an assessment and credentialing mechanism that is housed and managed online. Badges are designed to make visible and validate learning in both formal and informal settings." In education, many badges are digital and displayed on websites, in email signatures, or in online containers such as Mozilla Backpack, but the sky is the limit. Some organizations print out paper badges and place them outside of teachers' doors or issue badges as stickers that educators commonly place on laptops.

Microcredentialing may be considered a further application of badging, whereby the badges count towards an end goal, such as recertification hours. Ryerse (2017) defined it as "a digital certification indicating demonstrated competency in a specific skill ... educators apply their learnings in their practice, collect evidence, and demonstrate their competence."

Hodges et al. spoke to more uses of the practice:

> Badging and micro-credentialing may be one solution to the concerns about densely packed teacher preparation curricula. In addition to their use in initial teacher preparation, badges could be used to encourage

technology coordinators and in-service teachers to obtain credentials related to the evaluation of educational technology and the meaningful integration of those technologies once they are acquired. Professional learning networks could be important elements of this type of professional development. (2017, p. 21)

Professional Learning Networks

As we discussed in Chapter 2, social media has played an enormous role in shaping our society in recent years. Never before has it been so easy to communicate beyond borders. Our network no longer must be limited to only those educators within the four walls of our schools, with the occasional opportunity to interact with colleagues at district or state events; we can now literally learn from (and with) people located nearly everywhere. As Spirrison (2016) pointed out:

> Global education is predicated on the interconnectedness of educators who share values, ideas, and resources from their own communities and networks with the world. Technology and social media today provide teachers with access to an abundance of tools, voices, and instructional approaches that originate from a variety of cultures and societies.

Krutka and Carpenter (2016) identified several ways that educators use social media, specifically Twitter, such as finding resources, supporting new teachers, and facilitating professional learning opportunities, although it still remains untapped by the masses. Additionally, some educators struggle with setting a purpose, utilizing social media in the classroom, and even maintaining professionalism online.

Further, in reflecting upon the lack of cultural diversity among the respondents, the researchers asserted that "educators' uses of Twitter might simply reinforce privilege, segregation, or social inequity," referencing "homophily, or the tendency to seek out colleagues who are like minded or culturally similar" (Thelwall, 2009, in Krutka & Carpenter, 2016). They also warned of the possibilities of groupthink and echo chambers, hidden behind the term *likemindedness*. Although this can be the case, several PLNs, such as EduMatch (edumatch.org), do feature diversity of all kinds, from racial to national to religious, spanning every demographic. At the time of this writing, EduMatch has a global reach of over 30,000 educators, located on six continents, and connects them through all forms of social media. In an ISTE Ignite session, Thomas (2017a) cited the need for "deeper connections" where educators get to know one another as people, in order to have conversations on topics that may be considered more serious or polarizing. That way, they will be more likely to hear and reflect upon one another's points, even if they disagree, leading to potential growth.

If you are willing to embrace social media, the payoff can be immense for yourself as an instructor, your school or district, and most importantly, your students. (Even we, the authors of this book, met and developed a sisterhood through the power of social media!) According to Brad Spirrison of Participate, there are now "hundreds of self-organized Twitter chats to discuss best practices with peers" (2016) on topics ranging from prekindergarten (#ecechat) to higher education (#isteTEN, #teacheredchat) and everything in between. There are hashtags focused on specific topics mentioned in this book, such as digital equity (#digitalequity, #ISTEDEPLN), social justice (#equity, #educolor), and even educational technology (#edtech, #ISTE__ [insert year], #edtechchat). Krutka and Carpenter (2016) also reported:

> A large number of participants praised Twitter as a medium that empowered them to grow in their craft.

Participants also praised Twitter as the "best professional development," and "free and better than most paid workshops." To this latter point, a veteran teacher from Wisconsin elaborated that "it is free, and available when you want it."

Wagner (2018) expanded on yet another use of the PLN, stating "fortunately, with the development of AI [artificial intelligence] and global learning ecosystems, we have more control over and access to professional development opportunities." The author described virtual professional learning, powered by PLNs, such as Google's *Education on Air, The Global Education Conference,* EduMatch *Tweet & Talks,* and Edcamps such as *Edcamp Voice* and *Edcamp EduMatch,* which "provide a range of workshops and training for educators spread across a multitude of topics … you can also offer training on topics you feel most comfortable delivering."

However, Laat and Schreurs (2013) argued that "however powerful informal learning may be, there is a difficulty when it comes to utilizing it for professional development. Informal learning activities are mostly implicit, ad hoc, spontaneous, and invisible to others" (p. 2). Therefore, to make these connections more visible, they offer the idea of networked learning, an "emerging perspective that aims to understand learning processes by asking how people develop and maintain a web of social relations for their learning and professional development" (p. 4). This is yet another example of what can happen when PLNs and learning analytics/AI merge.

Learning Analytics—AI

It may seem like something out of a movie, but artificial intelligence (AI) is here, and making waves in the field of education. Possibly one of the most interesting examples to date was that of Jill Watson:

Earlier this year, students taking a course in Artificial Intelligence at Georgia Institute of Technology discovered that Jill Watson, one of their Teaching Assistants, was an AI creation. "Watson" was able to participate in online forum discussions, answering students' routine questions, posting reminders about deadlines, and introducing mid-week conversation topics to encourage students to share their thoughts. (Holroyd, 2016)

Who would have thought that a bot could be a teacher's assistant at a major university? Who would have ever imagined that, thanks to AI, we could now even automate blog posts? And, who would have guessed that a paragraph like the following could be written with the help of an AI tool, for free, online in minutes (AI-Writer; ai-writer.com):

Artificial intelligence (AI) is the next giant leap in learning and, according to those working in the field of education and technology, we haven't seen anything yet (Wood, 2016). As with virtual reality, artificial intelligence is still in its early days as an education tool, with minimal adoption in the classroom. But even more so than VR, artificial intelligence could fundamentally change the way we learn. (Roepke, 2017)

How do you even cite that? Granted, we did some basic editing, such as adding the in-text citations, although the website provided us with footnotes. Additionally, we edited the sequencing of the text, because the AI piecemealed the text verbatim from the sources. If we had known absolutely nothing about AI in education, however, we could have had a solid foundation for our research, all thanks to typing in the keywords *artificial intelligence, ed tech,* and *professional development.* The software did all of the research (for the preceding paragraph, not for the entire book).

Yet another site, Articoolo (articoolo.com), provided an even longer article, with tools for checking grammar, style, repetition, and structure. After some revision, the final output read:

> Artificial intelligence is predicted to change the nature of society by 2040. While futurist Ray Kurzweil claims AI will help us with the grand challenges facing humankind, Elon Musk warns us that artificial brains will be our greatest existential threat. Others argue that these have the potential for growth. Everything depends upon the way we handle the transition into this AI era. In 2016, the government under President Barack Obama introduced a nationwide strategic plan for artificial intelligence and, although not everyone welcomes this innovation, we still must remain current on how AI develops.
>
> Artificial Intelligence is transforming classes through customizable content, and also tracking and monitoring diagnostics. AI can automate basic routine work like grading simple tests. Pearson is working on a wide range of online games based on AI. Mindojo is developing algorithms which both teach and also study you to know precisely who you are.

Although Articoolo did not provide a reference list, other services allow the user to customize the results in styles such as APA and MLA. What are the implications for teaching and learning? Even more relevant to this section, what are the implications for professional development? Avery (2017) sums it up nicely by stating that, "AI can fill the gaps in subject areas in which a teacher doesn't have a particular expertise or help train teachers when there is a skill shortage in the job market, too." Again, with teachers as lifelong learners, AI and big data tools can help to personalize our own professional learning through analytics. LaPierre (2018) explained:

Similarly to how AI can provide personalized tutoring to students, the technology could offer ongoing professional development for teachers, too! These flexible, AI-driven courses could synthesize data on an educator's teaching experiences, abilities, and styles, then use this information to provide tailor-made trainings designed specifically to meet that individual's needs. Using web-enabled mobile devices like smartphones and tablets, educators would have the flexibility to complete these trainings at any time, no matter their present location—that's right: PD in your PJs.

Technologies such as these have already begun to emerge, such as the G Suite Training Chrome extension, "that offers simple and interactive training lessons to get you up and running fast with G Suite" (Google, n.d.).

Edcamps and Unconferences

Edcamps are an informal type of professional development where there are no presenters, just conversations about whatever the attendees want to learn or share. As Howard and Thomas (2016) explained:

> The Edcamp premise is unlike most forms of professional learning. Instead of traditional conferences, where organizers usually serve as gatekeepers and must approve which kind of knowledge can be provided by whom, Edcamps work differently. Conversations are the platform for the sharing of information.

Typically, the day begins with a brainstorm of topics on the session board, where people claim a room for a session to discuss the subject of interest. You never know who you'll be sitting next to at an Edcamp—perhaps a teacher, an administrator, a superintendent ... or maybe even a student, parent, or member of the community! The beauty of this model is that everyone has an equal voice and can share their questions, experiences, and ideas.

Preliminary findings from a recent study on teachers' professional learning preferences indicate that teachers feel more supported in their work and better prepared to support their students' learning when they select their own professional learning (PL) opportunities. Out of the educators surveyed, 55.4% indicate that the PL sessions they are required to attend are not relevant to their work. Nearly 64% report that these required PL sessions do not support them in their work. When asked about PL opportunities that they choose, 98% indicate that they feel more supported in their jobs and better prepared to support their students' learning. As if it were a surprise that educators feel more supported and prepared when they choose their own learning experiences, approximately 79% indicated that unconference PL opportunities are effective in general (Howard & Thomas, 2016). Furthermore, according to Hertz (2016), "what teachers want are the things that we always say we want to give our students: choice and voice."

Video and Livestreaming

Major players in the field of education are taking advantage of the power of video, with organizations such as ISTE, Discovery Education, and the Association for Supervision and Curriculum Development (ASCD) offering robust channels on YouTube or their own platforms. Video is not monopolized by only the power players, however, as evidenced by the rise in grassroots videos facilitated by the launch of YouTube in 2005.

Also, livestreaming is rapidly gaining popularity, utilized by content creators of all levels, to share what is occurring within a given moment. Generally, audience members can watch and interact through a device. Several livestreaming platforms are currently in existence, including YouTube itself, Periscope, Facebook Live, and many more. Facebook Live, for example, is a popular service, considering the popularity of the platform alone.

According to Erin*tegration (2016), "Facebook Live has the potential to directly impact professional development for teachers as teachers can stream tips, ideas, lesson demos and more directly to other teachers." Erin then proposed a variety of best practices for educators for using Facebook Live, including:

- Have a fully charged phone with strong Wi-Fi.
- Be ready.
- Know what you are going to say.
- Think about lighting.
- Be yourself.
- Engage in the conversation.
- Remind viewers who you are and where to find you online.
- Shake off mistakes, and move on.
- Don't flip the camera from front- to rear-facing while recording.

Although livestreaming is gaining in popularity, some educators and students remain hesitant to appear on camera. Therefore, Thomas (2018) proposed four more strategies to help readers "get used to the educator in the mirror," which included:

- **Practice.** Rehearse what you want to present. Not only will this help you get more comfortable with the words, you can also fine-tune your timing and emphasis.

- **Get eyes on it.** Solicit feedback from peers and students to ensure you're conveying the message you think you are.
- **Fire one and done.** If a rehearsed speech sounds too un-natural or simply isn't working for you, revert to the one-and-done approach of a livestream.
- **If all else fails, strip the video.** If you simply can't get the visual portion right or aren't comfortable on camera, separate the visual portion from the audio track during the editing and delete it. Keep the audio to use as a podcast.

Lewis (in Thomas et al., 2016) wrote of a grassroots movement called #PasstheScopeEDU, which utilizes the Periscope platform (owned by Twitter) in order to share stories from educators about what really happens within the profession. Initially, this was in response to rising frustration from teachers, concerning media gatekeepers promoting a predominantly negative narrative about educators and schools. As Lewis explained:

> Our Voxer community became a back channel and space where we could hash out ideas, vote on themed topics using DotStorm, perfect our skills and welcome others to come learn and do the same. To say this thing has gotten serious is not a joke! We were influencers that wanted to do things our way and share the love of learning with the world. We jumped all in and scoped about summer learning, the source of inspiration for our creative mojo, and new things we want to implement in the upcoming school year ... This helped us to realize how we could leverage this tool for professional and personalized learning for educators everywhere. You cannot get any more real or authentic than that. (p. 204)

At the time of this writing, #PasstheScopeEDU recently celebrated its second anniversary, which has evidenced how much educators want to connect and share. Continuing along this vein, Kaplinsky (2016) wrote, "A teacher who doesn't collaborate works on an isolated island. When this lack of collaboration permeates an entire school, teachers more closely resemble independent contractors than colleagues. I'm growing increasingly concerned that this is becoming more, and not less, common." As a solution, he created a movement called #ObserveMe, explaining "Observing other teachers in action has been a fabulous source of new ideas during my career. Similarly, the feedback I've received on my own teaching has given me perspective I didn't know was missing." Either online or in person, #ObserveMe allows for peer feedback about a particular lesson, to help the requestor improve. "One of the best ways to improve practice is to have colleagues observe one another and provide suggestions … We should welcome others' constructive feedback and practice giving it as well. Without it we aren't able to adjust our practice," Kaplinsky stated.

 Reflect & Consider

Videos and Student Privacy

As powerful as video feedback can be, there are some steps that are critical to teach preservice teachers as they consider this method in their future classrooms. Thomas (2015c) identified four steps to consider with regards to maintaining K–12 student privacy and security while livestreaming. Encourage preservice teachers to:

1. **Make sure their students have a publicity release on file.** Parents have the option to check whether or not they give permission for their child to be shown online or in any other medium. An affirmative response is typically mandatory for student athletes.

2. **Ensure their students are thirteen or older,** depending on the requirements for setting up an account on the site,

otherwise parent permission may also be required to use the livestreaming site.

3. **Confirm your students' consent to be livestreamed.** This is probably the most important of all, because it is basic human respect. Do not film those who do not want to be filmed, regardless of age. Period.

4. **Disable replays.** For example, when using Persicope immediately tap on the progress bar when it prepares for playback. This disables anyone from going back and watching it. However, auto-saving the Periscope livestream to your camera roll by clicking on the gear icon in the top-right corner of your profile page allows you to save and share the video later. *(Pro-tip: This may be very helpful for flipped instruction.)*

▲ ▲ ▲

Safety and privacy should be at the forefront of all that we do, as it is our professional responsibility. Thus, please keep these ideas in mind prior to sharing any information about students online, especially those under the age of 13.

Partnerships

Another strategy to prepare teachers with regard to digital equity would be to establish partnerships between PK–12 and institutions of higher education. North Carolina has worked on this, adding these partnerships to their state plan, which sought to incorporate school visits and collaboration between professors and schools.

In addition, this partnership between The William and Ida Friday Institute and the North Carolina Department of Education sought to share instructional materials with instructors in educator preparation programs, by granting them with licenses, so that they were well-informed of what awaited their teacher candidates. North Carolina also utilized cross-level microcredentialing opportunities for professional learning in technology.

According to organizers, "the overall goal of this effort is to update preparation programs so that superintendents, principals, experienced teachers and students report that new teachers and administrators are well prepared for their roles as digital-age educators" (William and Ida Friday Institute, 2015, p. 45–46). Other states could do well by following this model.

Partnerships with other organizations, such as grant awarders, may also be beneficial to educator preparation programs and schools alike. Several opportunities are available which can help to sustain digital equity initiatives, including Every Student Succeeds Act (ESSA) Title II funding (U.S. Department of Education, 2016). In fact, the U.S. Department of Education specifically recommends that state departments of education partner with higher education institutions for teacher residencies and alternative certification:

> For example, Teach Tomorrow in Oakland (TTO), developed by Oakland Unified School District, is a program that complements candidates receiving alternative certification through higher education teacher preparation programs by creating a pathway into teaching for new educators, most of whom are recruited locally and are people of color from minority groups that are underrepresented in the teaching profession. (p. 8)

Furthermore, specifically addressing equity, the Department of Education stated:

> To ensure that every student has access to excellent educators, the State Educational Agency (SEA) and

Local Educational Agency (LEA) must work together to develop, attract, and retain excellent educators in all schools, especially in high-need schools. Part of the purpose of the Title II, Part A program is to provide students from low-income families and minority students greater access to effective teachers, principals, and other school leaders (Elementary and Secondary Education Act [ESEA]). To realize this outcome, SEAs and LEAs are strongly encouraged to use Title II, Part A funds to improve equitable access to effective teachers. (p. 20)

States such as Illinois and Indiana have already documented their intent to collaborate with institutions of higher education to build capacity.

Finally, educational stakeholders may also be interested in partnering with such organizations as ISTE, which is beginning to offer certification for educators. "ISTE Certification for Educators is a new competency-based, vendor- and device-neutral teacher certification based on the ISTE Standards for Educators. This digital credential recognizes educators who understand how to use ed tech for learning in meaningful and transformative ways" (ISTE, 2018b). Educators can become

 DE Stories

Dr. Lara Kassab and Abigale Almerido's Perspectives

In 2016, California adopted new Teacher Performance Expectation standards (TPEs) for credentialing new teachers. The TPEs integrated a more specific and holistic view of technology as compared to the 2013 TPEs. The TPEs do not specifically call for an understanding and addressing of digital equity, instead they require teacher candidates to "demonstrate knowledge of effective teaching strategies aligned with the internationally recognized educational technology standards" (such as ISTE). Our teacher preparation program [at San Jose State University], however, has a reputation for a commitment to equity, social justice, and social-emotional learning (SEL). The equity piece of pursuing digital equity was a natural to incorporate, but integrating the technology component was the challenge.

As I, and my colleagues at San Jose State University began to explore online options for introducing teacher candidates to authentic uses of technology into our program, we reached out to our local and county offices of education for details on their expectations for new teachers. Additionally, I posted requests for ideas on how new teachers should be prepared for technology use from my K–12 teaching friends in social media. A former teacher candidate, Abigale, saw this request and reached out to me about creating a partnership. She had been teaching for several years and had recently become a technology specialist for the Santa Clara County Office of Education (SCCOE).

In her initial outreach, Abby provided the feedback that teacher candidates, upon entering the profession, lacked educational technology skills. Based on this feedback, she and her colleagues at the SCCOE have met with both the teacher education faculty and college leadership to discuss potential partnership activities. Together, we are exploring the use of the SCCOE office's Professional Innovation Pathways (innovation.sccoe.org) online courses for faculty and candidates. The academies and courses could support equitable access to knowledge for teaching with technology. They are freely available to county educators, regardless of district or school funding. The flexibility of online platforms allows anytime, anywhere, self-paced, and self-selected development in various topics, which can meet participants where they are. Abigale and her colleagues have also offered to provide ongoing face-to-face professional development for faculty so that a commitment to equity can expand to include digital as well as social, emotional, and academic aspects.

Abigale is currently working on developing an Academy about culturally responsive teaching and leading with equity. The courses I teach at San Jose State University are focused on culturally responsive teaching and ensuring equity for K–12 students and teachers. Abigale has asked for us to work together on the concepts for the course. This allows me the opportunity to provide expertise in these areas while concurrently developing ideas of how technology and digital equity could be woven into my courses. With models in hand, faculty conversations around digital aspects of commitments to equity could be less daunting.

We have started a strong partnership, in which Abigale and her colleagues have brought "state-of-the-state" of local K–12 technology use and ideas to faculty, and I and my colleagues are bringing our expertise in equity to the county. This partnership highlights the importance of building and maintaining relationships between teacher education programs, district leadership, local teachers, and former students. I hope that we will continue the partnership, and co-publish the work we are doing as a model of how, working together, we can create strong teachers committed to academic, social, emotional, and digital equity for all students.

certified by attending training, which lasts approximately thirty hours, then submitting a portfolio of artifacts. More information will be shared on the ISTE website as the program continues to develop.

What Would You Do?

Imagine teaching a new course that has traditionally used paper and pencil only. This course is not a technology course, yet it is a course that preservice teachers are required to take in order to obtain their credential. You've been asked to incorporate technology. *What would you do?* How would you decide which emerging technologies to integrate without knowing the level of your students' technology knowledge? Would you first visit local K–12 schools to find out what they are using on their campuses?

You might begin by seeking the input of your preservice teachers to identify their interests and tech-knowledge. Most importantly, you should make sure the emerging technology is integrated seamlessly and not an add-on.

Join the conversation on Twitter using #DigEquityBook, and tell us what you would do.

DE Wisdom

While I do agree that there are certain elements of technology that teachers are sometimes unaware of, I also believe that, if we are asking teachers to serve students according to their needs, we must engage teachers and serve their needs the same.

—*Rafranz Davis, 2015*

Never forget to meet your learning teachers where they are—make no assumptions, only observations, about their knowledge and skill level. Find out what they want to learn and how they learn best, before diving in to teach new innovations. As you enact digital equity strategies in university classrooms, you are modeling effective approaches that will inevitably be adopted by your preservice teachers.

If you have tips on how to directly meet the needs of learners in your classroom through digital equity strategies, please share them on Twitter using #DigEquityBook.

CONCLUSION

We realize that preparing preservice teachers for utilizing emerging technologies in their future classrooms has had its challenges for years and that the inclusion of digital equity has not always been a priority. Research and stories discussed throughout this book indicate that professors and professional development leaders can make a huge impact on preservice teacher learning; therefore, the burden to facilitate meaningful digital equity instruction rests on their shoulders.

We propose that now is the time for professors to be learners alongside their students in order to be best prepared to take on this work. Krueger and James (2017) suggested the right to digital equity ensures the right to connect to needed resources—anywhere, anytime. Clearly, the work is never-ending as technology will continue to evolve, but now is the time for pioneering educators to do work towards closing the digital equity gap.

We'd like to offer one final takeaway in Table C.1, that you may wish to consider using in your teacher preparation course. You might use it as is or simply offer your preservice teachers a blank chart that you complete together, followed by a rich discussion about digital equity practices. This table may also serve as a reminder or visual reference for preservice teachers to return to once they begin working with their own future students.

Conclusion

Table C.1 Digital Equity Versus Digital Inequity in Teacher Preparation

Digital Equity	Digital Inequity
Modeling the tech tool selection process (e.g. pedagogy, diversity, usability, privacy)	Choosing a tool simply because it's popular and implementing it without context
Considering representation	Settling for tokenism
Providing each student with an opportunity for active learning	Allowing opportunities for creation only in STEM, methods, technology-based, or other higher-level courses
Connecting with classes globally to learn together, based on mutual respect and goals	Connecting with classes globally to "save" the other party or to "teach" them without also learning from (and with) them
Finding out which students are comfortable using social media for class discussions and offering an alternative option (e.g. discussion board) for those who are not	Requiring all students to use their social media accounts for class discussions, regardless of their comfort level or personal/professional preferences
Ensuring a tool is culturally appropriate and inclusive of non-offensive imagery and current language (e.g., refers to enslaved peoples instead of slaves)	Using a tool regardless of harmful language and offensive imagery (e.g., certain "slave" simulations)
Assessing the numbers of devices in your class to determine effective grouping strategies (e.g. 3:1 or 4:1)	Requiring all students have a device in class without consideration given to financial barriers

REFERENCES

Adichie, C. N. (2009). *Chimamanda Adichie: The danger of a single story* [Video file]. Retrieved from https://www.ted.com/talks/chimamanda_adichie_the_danger_of_a_single_story

Albee, J. J. (2003). A study of preservice elementary teachers' technology skill preparedness and examples of how it can be increased. *Journal of Technology and Teacher Education, 11*(1), 53–71.

Alliance for Excellence in Education. (2016). *Digital equity: The road ahead* [Video file]. Retrieved from https://digitallearningday.org/dld-2016/digital-equity-the-road-ahead

Avery, A. (2017). Artificial intelligence promises a personalized education for all. *The Possibility Report.* Retrieved from https://www.theatlantic.com/sponsored/vmware-2017/personalized-education/1667

Beaumont, C. (2008, December 16). Technology review of the year 2008. *The Telegraph.* Retrieved from https://www.telegraph.co.uk/technology/3793173/Technology-review-of-the-year-2008.html

Borrero, N. E., Flores, E., & de la Cruz, G. (2016). Developing and enacting culturally relevant pedagogy: Voices of new teachers of color. *Equity & Excellence In Education, 49*(1), 27–40.

Bradshaw, J., Clegg, S., & Trayburn, D. (1995). An investigation into gender bias in educational software used in English primary schools. *Gender and Education, 7*(2), 167–174.

References

Bull, G., Spector, J. M., Persichitte, K., & Meiers, E. (2017). Reflections on preparing educators to evaluate the efficacy of educational technology: An interview with Joseph South. *Contemporary Issues in Technology and Teacher Education, 17*(1). Retrieved from www.citejournal.org/volume-17/issue-1-17/editorial/reflections-on-preparing-educators-to-evaluate-the-efficacy-of-educational-technology-an-interview-with-joseph-south

Burton, L. (2016). ISTE student standards promote equity in education [Blog post]. Retrieved from https://www.iste.org/explore/articleDetail?articleid=818

Cable News Network. (2004, December 17). *Year of innovations, challenges*. Retrieved from www.cnn.com/2004/TECH/12/17/yir.scitech

Chen, C.-H. (2007). Cultural diversity in instructional design for technology-based education. *British Journal of Educational Technology, 38*(6), 1113–1116.

Computer Hope. (n.d.). *Computer history for the year 2005*. Retrieved from https://www.computerhope.com/history/2005.htm

Consortium for School Networking. (2016). *CoSN's 2016 annual infrastructure survey*. Retrieved from https://cosn.org/sites/default/files/CoSN_4th_Annual_Survey_Nov 2 FINAL.pdf

Cooke, M. (2012). Twitter and Canadian educators. *Education Canada, 52*(2).

Cortez, M. B. (2017, April 17). Q&A: Richard Culatta aims to use government experience as new ISTE CEO. *EdTech: Focus on K–12*. Retrieved from https://edtechmagazine.com/k12/article/2017/04/qa-richard-culatta-aims-use-government-experience-new-iste-ceo

Cox, A. (2009). Visual representations of gender and computing in consumer and professional magazines. *New Technology, Work & Employment, 24*(1), 89–106.

Curran, M. B. (2012). The *iCitizen project*. Retrieved from https://www.mbfxc.com/the-icitizen-project.html

Darling-Hammond, L., Meyerson, D., LaPointe, M., & Orr, M. T. (2009). *Preparing principals for a changing world: Lessons from effective school leadership programs.* San Francisco, CA: Jossey-Bass.

Davies, D. (2018, January 10). *Looking back at 2018 in search: A time traveler's year in review.* Retrieved from https://searchengineland.com/2018-year-review-289384

Davis, R. (2015). *The missing voices in edtech: Bringing diversity into edtech.* Thousand Oaks, CA: Corwin.

Davis, T., Fuller, M., Jackson, S., Pittman, J., & Sweet, J. (2007). *A national consideration of digital equity.* Washington, DC: International Society for Technology in Education.

Dieker, L., Kennedy, M. J., Smith, S., Vasquez, E., III, Rock, M., & Thomas, C. (2014). Use of technology in the preparation of pre-service teachers (Document No. IC-11). Retrieved from ceedar.education.ufl.edu/wp-content/uploads/2014/10/IC-11_FINAL_05-26-15.pdf

Dogtiev, A. (2018). *Uber revenue and usage statistics* (2017). Retrieved from www.businessofapps.com/data/uber-statistics

Durden, T., Dooley, C. M., & Truscott, D. (2016). Race still matters: Preparing culturally relevant teachers. *Race Ethnicity and Education, 19*(5), 1003–1024.

Eng, J. (2015, December 23). *Your data at risk: 2015 was a year full of memorable hacks.* Retrieved from https://www.nbcnews.com/storyline/2015-year-in-review/your-data-risk-2015-was-year-full-memorable-hacks-n474656

References

Erin*tegration. (2016, May 15). *Facebook Live for teachers.* Retrieved from https://www.erintegration.com/2016/05/15/facebook-live-teachers

Ertmer, P. A., & Ottenbreit-Leftwich, A. T. (2010). Teacher technology change: How knowledge, confidence, beliefs, and culture intersect. *Journal of Research on Technology in Education, 42*(3), 255–284.

Ford, M. (2017, July 7). A major victory for the right to record police. *The Atlantic.* Retrieved from https://www.theatlantic.com/politics/archive/2017/07/a-major-victory-for-the-right-to-record-police/533031

Gallardo-Echenique, E. E., Marqués-Molías, L., Bullen, M., & Strijbos, J.-W. (2015). Let's talk about digital learners in the digital era. *International Review of Research in Open and Distributed Learning, 16*(3), 156–187.

Google. (n.d.). *G Suite training.* Retrieved from https://chrome.google.com/webstore/detail/g-suite-training/idkloemkmldbemijiamdiolojbffnjlh?hl=en

Google Innovator Program. (2017, March 28). *Innovation live: Advocate for school change* [Video file]. Retrieved from https://www.youtube.com/watch?v=PQYueLd_xHw

Graham, J. (2017, December 7). These were the most downloaded Apple apps of 2017. *USA Today.* Retrieved from https://www.usatoday.com/story/tech/talkingtech/2017/12/07/most-downloaded-apple-apps-2017/928745001

Gross, D. (2009, December 23). #%*@#! *The top 10 tech 'fails' of 2009.* Retrieved from www.cnn.com/2009/TECH/12/23/tech.fail/index.html

Gross, D. (2010, December 27). *The top 10 tech trends of 2010.* Retrieved from www.cnn.com/2010/TECH/innovation/12/27/top.tech.trends.year/index.html

Grossman, L. (2010, December 15). Person of the year 2010: Mark Zuckerberg. *Time*. Retrieved from content.time.com/time/specials/packages/article/0,28804,2036683_2037183,00.html

Hamburger, E. (2013, March 1). The age of the brag is over: Why Facebook might be losing teens. *The Verge*. Retrieved from https://www.theverge.com/2013/3/1/4049592/the-age-of-the-brag-is-over-why-facebook-might-be-losing-teens

Hamburger, E. (2013, August 22). Facebook isn't making you depressed, but the internet is. *The Verge*. Retrieved from https://www.theverge.com/2013/8/22/4647916/facebook-isnt-making-you-depressed-the-internet-is

Hanson, K. (1997). Gender, "discourse," and technology. Center for Equity and Diversity, working paper 5. Newton, MA: Education Development Center.

Hertz, M. B. (2016, April 15). Why edcamp is the future of PD [Blog post]. *Edutopia: A George Lucas Educational Foundation*. Retrieved from https://www.edutopia.org/blog/edcamp-is-future-of-pd-mary-beth-hertz

Hillebrand, A. (2018, April 8). Physical education instructor receives grant for heart rate monitors. *The Guidon*. Retrieved from https://hayshighguidon.com/news/2018/04/08/physical-education-instructor-receives-grant-for-heart-rate-monitors

Hodges, C. B., Carpenter, J. P., & Borthwick, A. C. (2017). Commentary: Response of the American Association of Colleges for Teacher Education to "An interview with Joseph South" regarding the preparation of educators to evaluate the efficacy of educational technology. *Contemporary Issues in Technology and Teacher Education, 17*(1), 17–23.

References

Holroyd, L. (2016, September 29). Into the unknown: Professional development for future educators [Blog post]. Retrieved from https://blog.oup.com/2016/09/professional-development-future-educators

Holt, C. B., & Garcia, P. (n.d.). *Preparing teachers for children in poverty*. Retrieved from www.aasa.org/SchoolAdministratorArticle.aspx?id=7576

Howard, K. E. (2013). Using Facebook and other SNSs in K–12 classrooms: Ethical considerations for safe social networking. *Issues in Teacher Education, 22*(2), 39–54.

Howard, K. E., & Howard, N. R. (2014, July). 8 considerations for social networks in classrooms. Retrieved from https://www.ecampusnews.com/featured/featured-best-practice/classrooms-social-networks-362

Howard, K. E., Sauceda Curwen, M., Howard, N. R., & Colón-Muñiz, A. (2015). Attitudes toward using social networking sites in educational settings with underperforming Latino youth: A mixed methods study. *Urban Education. 50*(8), 989–1018.

Howard, N. R. (2012, July 18). I'm not your friend: Social networking in university classes. *Educause Review*. Retrieved from www.educause.edu/ero/article/im-not-your-friend-social-networking-university-classes

Howard, N. R. (2015, October 23). 5 "BEs" for connected and curious educators [Blog post]. *Edutopia: A George Lucas Educational Foundation*. Retrieved from https://www.edutopia.org/blog/5-bes-connected-curious-educators-nicol-howard

Howard, N. R., & Dabbs, L. (2015). *Standing in the gap: Empowering new teachers through connected resources*. Thousand Oaks, CA: Corwin.

Howard, N. R., & Thomas, S. J. (2016, July 20). EdCamps: The new professional development [Blog post]. *Edutopia: A George Lucas Educational Foundation*. Retrieved from https://www.edutopia.org/blog/edcamps-the-new-professional-development-nicol-howard-sarah-thomas

Howard, T. C., Flennaugh, T. K., & Terry, Sr., C. L. (2012). Black males, social imagery, and the disruption of pathological identities: Implications for research and teaching. *Educational Foundations, 26*(1/2), 85–102.

Hsu, P. S. (2012). Examining the impact of educational technology courses on pre-service teachers' development of technological pedagogical content knowledge. *Teaching Education, 23*(2), 195–213.

Hughes, J. E., Liu, S., & Lim, M. (2016). Technological modeling: Faculty use of technologies in preservice teacher education from 2004 to 2012. *Contemporary Issues in Technology and Teacher Education, 16*(2), 184–207.

International Society for Technology in Education. (2017a). *Five things every educator should know about digital equity.* Retrieved from https://www.iste.org/explore/articleDetail?articleid=654

International Society for Technology in Education. (2017b). *ISTE Standards for Educators.* Retrieved from https://www.iste.org/standards/for-educators

International Society for Technology in Education. (2018). *ISTE Certification.* Retrieved from https://www.iste.org/learn/iste-certification

Johnson, B. (2009, December 28). Facebook more than doubled its number of users this year. Retrieved from https://www.theguardian.com/technology/2009/dec/28/facebook-users-social-network

References

Johnson, C. J. (2017, December 21). The year in Trump tweets. *Chicago Tribune.* Retrieved from www.chicagotribune.com/news/yearinreview/ct-the-year-in-trump-tweets-20171215-story.html

Jones, M. J., & Dexter, S. S. (2018). Teacher perspectives on technology integration professional development: Formal, informal, and independent learning activities. *Journal of Educational Multimedia & Hypermedia, 27*(1), 83–102.

Kaplinsky, R. (2016, August 15). *#Observeme.* Retrieved from https://robertkaplinsky.com/observeme

Karl, K., & Peluchette, J. V. (2011). "Friending" professors, parents and bosses: A Facebook connection conundrum, *Journal of Education for Business, 86*(4), 214–222.

Keneally, M. (2015, December 29). *Year in review: 13 biggest news stories of 2015.* Retrieved from https://abcnews.go.com/US/year-review-13-biggest-news-stories-2015/story?id=35852690

Krueger, K., & James, J. (2017, March/April). Digital equity: The civil rights issue of our time. *Principal.* Retrieved from https://www.naesp.org/principal-marchapril-2017-technology-all/digital-equity-civil-rights-issue-our-time

Krutka, D. G., & Carpenter, J. P. (2016). Participatory learning through social media: How and why social studies educators use Twitter. *Contemporary Issues in Technology and Teacher Education, 16*(1). Retrieved from www.citejournal.org/volume-16/issue-1-16/social-studies/participatory-learning-through-social-media-how-and-why-social-studies-educators-use-twitter

Kruzman, D. (2017, May 26). Protests, creepy clowns and the mannequin challenge: The 2016–17 college year in review. *USA Today.* Retrieved from http://college.usatoday.com/2017/05/26/2016-2017-school-year-in-review/

Laat, M., & Schreurs, B. (2013). Visualizing informal professional development networks: Building a case for learning analytics in the workplace. *American Behavioral Scientist, 57*(10), 1421–1438.

Langone, A. (2018, March 22). #Metoo and time's up founders explain the difference between the 2 movements—and how they're alike. *Time*. Retrieved from time.com/5189945/whats-the-difference-between-the-metoo-and-times-up-movements

LaPierre, J. (2018, March 12). How AI supports teachers [Blog post]. Retrieved from https://www.filamentgames.com/blog/how-ai-supports-teachers

Loewus, L. (2017, September 5). How virtual reality is helping train new teachers. *Education Week*. Retrieved from https://www.edweek.org/ew/articles/2017/09/06/student-teachers-get-real-world-practice-via.html

Lynn, M. (2014, March 19). Making culturally relevant pedagogy relevant to aspiring teachers. *Diverse*. Retrieved from diverseeducation.com/article/61280

MacArthur Foundation. (n.d.). *Digital badges*. Retrieved from https://www.macfound.org/programs/digital-badges

Madrigal, A. C. (2011, December 6). The 12 most important tech stories of the year. *The Atlantic*. Retrieved from https://www.theatlantic.com/technology/archive/2011/12/the-12-most-important-tech-stories-of-the-year/249558/

Maguth, B. M., & Yamaguchi, M. (2013). The use of social networks in the social studies for global citizenship education: Reflecting on the March 11, 2011 disaster in Japan. *The Georgia Social Studies Journal, 3*(2), 80–93.

Mills, M. (2014). Effect of faculty member's use of Twitter as informal professional development during a preservice teacher internship. *Contemporary Issues in Technology and Teacher Education, 14*(4), 451–467. Retrieved from www.citejournal.org/vol14/iss4/currentpractice/article1.cfm

Molnar, M. (2014, November 13). Richard Culatta: Five ways technology can close equity gaps. *Education Week.* Retrieved from https://marketbrief.edweek.org/marketplace-k-12/richard_culatta_five_ways_technology_can_close_equity_gaps

Newcomb, A. (2016, December 27). 2016's tech obituaries: The year we lost Vine, Meerkat, and the fiery Note 7. Retrieved from https://www.nbcnews.com/storyline/2016-year-in-review/2016-s-tech-obituaries-year-we-lost-vine-meerkat-fiery-n700496

New Media Consortium. (2006). Horizon report 2006 higher ed edition. Retrieved from https://www.nmc.org/publication/nmc-horizon-report-2006-higher-ed-edition

New Media Consortium. (n.d.). *NMC Horizon.* Retrieved from https://www.nmc.org/nmc-horizon

Oculus Rift. (n.d.). *Step into Rift.* Retrieved from https://www.oculus.com/rift/#oui-csl-rift-games=star-trek

Overbay, A., Patterson, A., Vasu, E., & Grable, L. (2010). Constructivism and technology use: Findings from the IMPACTing leadership project. *Educational Media International, 47*(2), 103–120.

Paris, D. (2012). Culturally sustaining pedagogy: A needed change in stance, terminology, and practice. *Educational Researcher, 41*(3), 93–97.

Phillips, S. (2007, July 25). A brief history of Facebook. *The Guardian*. Retrieved from https://www.theguardian.com/technology/2007/jul/25/media.newmedia

Popper, B. (2014, November 26). Here's why GoPro wants to make its own drones. *The Verge*. Retrieved from https://www.theverge.com/2014/11/26/7295731/heres-why-gopro-wants-to-make-its-own-drones

Prensky, M. (2001). Digital natives, digital immigrants. *On the Horizon, 9*(5), 1–6.

Prensky, M. (2011). Is the digital native a myth? No. *Learning & Leading with Technology, 39*(3), 6–7.

Project Tomorrow. (2017). 2017 digital learning report from Blackboard and Speak Up. Retrieved from www.tomorrow.org/speakup/speak-up-2016-trends-digital-learning-june-2017.html

Resta, P., & Laferrière, T. (2015). Digital equity and intercultural education. *Education and Information Technologies, 20*(4), 743–756.

Riccards, P. (2016, October 13). Learning to teach. *US News & World Report*. Retrieved from https://www.usnews.com/opinion/knowledge-bank/articles/2016-10-13/imagine-the-teacher-education-of-the-future

Richtel, M. (2012, May 29). Wasting time is new divide in digital era. *The New York Times*. www.nytimes.com/2012/05/30/us/new-digital-divide-seen-in-wasting-time-online.html?pagewanted=all&_r=0

Rideout, V., & Katz, V. S. (2016). Opportunity for all? Technology and learning in lower-income families. Retrieved from digitalequityforlearning.org/wp-content/uploads/2015/12/jgcc_opportunityforall.pdf

References

Roberts, J. J. (2015, July 29). Man arrested for shooting $1,800 drone won't apologize, cites privacy. *Fortune*. Retrieved from fortune.com/2015/07/29/shoot-drone-privacy

Roblyer, M. D., McDaniel, M., Webb, M., Herman, J., & Witty, J. (2010). Findings on Facebook in higher education: A comparison of college faculty and student uses and perceptions of social networking sites. *The Internet and Higher Education, 13*(3), 134–140.

Roepke, J. (2017, April 24). How artificial intelligence will transform education. *EdTech Digest*. Retrieved from https://edtechdigest.com/2017/04/24/how-artificial-intelligence-will-transform-education

Rogers, E. M. (1962). *Diffusion of innovations*. East Lansing, MI: Dept. of Communication, Michigan State University.

Ryerse, M. (2017, November 7). Competency-based micro-credentials are transforming professional learning. *Getting Smart*. Retrieved from www.gettingsmart.com/2017/11/micro-credentials-transforming-professional-learning

Sanders, J. (2005). Gender and technology in education: A research review. In C. Skelton, B. Francis, & L. Smulyan (Eds.), *Handbook of Gender in Education*. London: Sage Publications.

Sheldon, J. P. (2004). Gender stereotypes in educational software for young children. *Sex Roles, 51*(7-8), 433–444.

Smart, A., & Corn, J. (2017). *Equity for digital-age learning: Education leaders address new challenges in North Carolina schools*. Retrieved from https://ncdli.fi.ncsu.edu/resources/docs/equity.pdf

Smiley, C., & Fakunle, D. (2016). From "brute" to "thug:" The demonization and criminalization of unarmed Black male victims in America. *Journal Of Human Behavior in the Social Environment, 26*(34), 350–366. Retrieved from doi.org/10.1080/10911359.2015.1129256

Solomon, G. (2002). Digital equity: It's not just about access anymore. *Technology & Learning, 22*(9).

Spirrison, B. (2016, June 2). The future of professional development is collaborative development. *Getting Smart.* Retrieved from www.gettingsmart.com/2016/06/future-professional-development-collaborative-development

Sutter, J. D. (2009, December 23). *The top 10 tech trends of 2009.* Retrieved from www.cnn.com/2009/TECH/12/22/top.tech.trends.2009/index.html

Thomas, S. J. (2015, February 1). Social justice and the new fourth estate. Retrieved from https://sarahdateechur.wordpress.com/2015/02/01/social-justice-and-the-new-fourth-estate

Thomas, S. J. (2015, May). Flipping: Five questions to ask about equity. Retrieved from https://tinyurl.com/fivequestionsonequity

Thomas, S. J. (2015, May 07). Student privacy on #PeriscopeEDU. Retrieved https://sarahdateechur.wordpress.com/2015/05/07/student-privacy-on-periscopeedu

Thomas, S. J., Simmons, T., Poth, R. D., Pierson, R. J., Harris, B., Ward, R., et al. (2016). *#Edumatch snapshot in education.* Alexandria, VA: EduMatch.

Thomas, S. J. (2017, April 2). Random thoughts on a Sunday morning (culture and personalization). Retrieved from https://sarahdateechur.wordpress.com/2017/04/02/random-thoughts-on-a-sunday-morning-the-interplay-of-personalization-and-culture

Thomas, S. J. (2018, January 16). *4 ways to get used to the educator in the mirror.* Retrieved from www.insightadvance.com/blog/4-ways-to-get-used-to-educator-in-the-mirror

References

United Nations. (2015). *Sustainable development goals.* Retrieved from https://www.un.org/sustainabledevelopment/sustainable-development-goals

U.S. Department of Education, National Center for Education Statistics. (2009). Table 427: Percent of home computer users using specific applications, by selected characteristics: 1997 and 2003. In U.S. Department of Education, National Center for Education Statistics (Ed.), *Digest of Education Statistics* (2009 ed.). Retrieved from https://nces.ed.gov/programs/digest/d03/tables/dt427.asp

U.S. Department of Education, National Center for Education Statistics. (2016). Table 702.10: Percentage of children ages 3 to 18 living in households with a computer, by type of computer and selected child and family characteristics: Selected years, 2010 through 2015. In U.S. Department of Education, National Center for Education Statistics (Ed.), *Digest of Education Statistics* (2016 ed.). Retrieved from https://nces.ed.gov/programs/digest/d16/tables/dt16_702.10.asp

U.S. Department of Education, Office for Civil Rights. (2014). *Dear colleague letter from the Assistant Secretary.* Washington, D.C.: U.S. Department of Education, Office for Civil Rights.

U.S. Department of Education, Office of Educational Technology. (2010). *Transforming American education: Learning powered by technology: Draft.* Washington D.C.: U.S. Department of Education, Office of Educational Technology.

U.S. Department of Education. (2016). Non-Regulatory Guidance for Title II, Part A: Building Systems of Support for Excellent Teaching and Leading. Washington D.C.: U.S. Department of Education.

U.S. Department of Education, Office of Educational Technology. (2016). Future ready learning: Reimagining the role of technology in education: National Education Technology Plan. Washington D.C.: U.S. Department of Education, Office of Educational Technology. Retrieved from https://tech.ed.gov/files/2015/12/NETP16.pdf

U.S. Department of Education, Office of Educational Technology. (2017). Reimagining the role of technology in education: National Education Technology Plan Update. Washington D.C.: U.S. Department of Education, Office of Educational Technology. Retrieved from https://tech.ed.gov/files/2017/01/NETP17.pdf

Verge. (2012, December 31). The year in review: The biggest stories of 2012. *The Verge*. Retrieved from https://www.theverge.com/2012/12/31/3813168/verge-year-in-review-2012

Verge. (2014, December 16). The 2014 Year in Review. *The Verge*. Retrieved from https://www.theverge.com/2014/12/16/7395527/the-year-in-review-the-biggest-stories-of-2014

Voithofer, R. (2009). Pre-service teachers' conceptions of the digital divide: A 4-year study at a predominantly white institution. Presented at the American Educational Research Association (AERA) National Conference, San Diego, CA.

Wagner, K. (2018, January 15). A blended environment: The future of AI and education. *Getting Smart*. Retrieved from www.gettingsmart.com/2018/01/a-blended-environment-the-future-of-ai-and-education

Wang, S.-K., Hsu, H.-Y., Campbell, T., Coster, D., & Longhurst, M. (2014). An investigation of middle school science teachers and students use of technology inside and outside of classrooms: Considering whether digital natives are more technology savvy than their teachers. *Educational Technology Research & Development, 62*(6), 637–662.

References

Warschauer, M. (2002, July). Reconceptualizing the digital divide. *First Monday, 7*(7).

Warschauer, M. (2007). A teacher's place in the digital divide. *Yearbook of the National Society for the Study of Education, 106*(2), 147–166.

Warschauer, M., & Matuchniak, T. (2010). New technology and digital worlds: Analyzing evidence of equity in access, use, and outcomes. *Review of Research in Education, 34*(1), 179–225

William and Ida Friday Institute for Educational Innovation. (2015). *North Carolina digital learning plan.* Retrieved from https://ncdli.fi.ncsu.edu/dlplan/docs/dlplan.pdf

Wood, A. (2016, September 28). Artificial intelligence is the next giant leap in education. *Raconteur.* Retrieved from https://www.raconteur.net/technology/artificial-intelligence-is-the-next-giant-leap-in-education

INDEX

A
access to internet/devices, 18–24
"Achieving Digital Equity in Teacher Education" project, 27
adaptive learning technologies, 53
advocate role, 91–92
AI-Writer, 100
Alliance for Excellence in Education, 10
Almerido, Abigale, 110–111
analytics technologies, 53, 99–102
apps, 37, 73
Articoolo, 101
artificial intelligence, 99–102
assumptions about teacher tech knowledge, 74–75
augmented reality, 56–58
Avery, A., 101

B
badging, 55–56, 96–97
barriers, overcoming
    about, 69
    assumptions about teacher tech knowledge, 74–75
    diversity and digital equity, 70–73
    gender differences, 72, 73
    quality teaching, 75–80
    racial bias, 73
    staying current, connected, and curious, 80–87
Beaverton School District, Oregon, 12–15
best practices, modeling, 10–11
blog posts, automated, 100–101
Borthwick, A. C., 90, 91, 96–97
Bradshaw, J., 72
"bring your own device" (BYOD) programs, 36
broadcasting, personal, 43–44

Brown, Michael, 39
Brown, Patricia, 70–71
buses, extended-activity, 12–13

C
California Teacher Performance Expectation standards, 109–110
Carpenter, J. P., 90, 91, 96–97, 98–99
certification, alternative, 108
Chapman University, 59
Clegg, S., 72
CNN, 37
collaboration lesson example, 17, 19
Colón-Muñiz, A., 72
connected, staying, 80–87
Cooke, M., 82
Culatta, Richard, 6, 16
cultural background of students, 28–29, 91–92
culturally relevant pedagogy (CRP), 91, 92
culturally sustaining pedagogy (CSP), 91–92
curiosity, 80–87
Curran, Marialice, 82–85
current, staying, 80–87

D
Davis, Rafranz, 70, 112
DE Stories
    Curran, Marialice, 82–85
    Hiefield, Matthew, 12–15
    Howard, Nicol R., 41–43
    Hsieh, Betina, 78–79
    Kassab, Lara, 109–111
    Quan-Lorey, Stephanie, 60–63

Index

DE Wisdom
  cultural background of students, 28–29
  needs of teachers, 112–113
  social media, 45–47
  staying current, connected, and curious, 86–87
  technology as tool, 66–67
device/internet access, 18–24
digital citizens, 83–85
digital divide, 19–23
digital equity. *See also specific topics*
  defined, 1, 2, 5, 15–16
  digital divide *versus*, 23
  digital inequity *versus*, 116
digital immigrants, 93–95
digital inequity, 116
digital natives, 93–95
diversity
  about, 70
  gender differences, 72, 73
  open letter, 70–71
  racial bias, 73

E

Edcamps, 99, 102–103
education spheres, interconnections of, 64
educational gaming, 55–56
educator preparation. *See* teacher preparation, critical issues in; teacher preparation, non-traditional; teacher preparation, traditional
educators. *See* teachers
EDUCAUSE, 53
EduMatch, 98, 99
8 Considerations for Social Networks in Classrooms (Howard & Howard), 39–40
E-Rate, 20
Erin*tegration, 104
ethnicity and device access, 20
Every Student Succeeds Act Title II funding, 108–109

F

Facebook, 41–42. *See also* social media
Facebook Live, 104
Ferguson, Missouri, 39
flipped learning, 8, 51–52
Flores, Nirmla, 26–27
Ford, James, 7

G

G Suite Training Chrome extension, 102
game-based learning, 55–56
gender differences, 72, 73
Georgia Institute of Technology, 100
grassroots video, 43–44, 64

H

hashtags, 46
Hertz, M. B., 103
Hiefield, Matthew, 12–15
Higher Education Report, 54
Hodges, C. B., 90, 91, 96–97
home technology use overview, 7–9
Horizon Report
  about, 32–33
  augmented reality and virtual reality, 57
  educational gaming, 55–56
  makerspaces, 58–59, 60
  mobile devices, 35–36, 37
  online learning, 50, 51, 53, 54
  personal broadcasting, 43–44
  social media, 38–39
  technology adoption predictions, 33–35, 63–65
hotspots, 13, 14, 21
Howard, Keith E., 39–40, 45–46, 69, 72, 93–95
Howard, Nicol R., 39–40, 41–43, 45–46, 69, 72, 81, 86–87, 102
Hsieh, Betina, 78–79
Hughes, J. E., 43, 49

Index

I
iCitizen Project, 84–85
iMentor research, 82–83
implementation plans, 24–25
information, critical consumers of, 90–91
internet/device access, 18–24
ISTE Certification for Educators, 109, 112
ISTE Standards for Educators, 3–4, 73, 77, 78–79
ISTE Standards for Students, 78–79

J
James, J., 115

K
Kajeet hotspot project, 14
Kaplinsky, R., 106
Kassab, Lara, 109–111
Katz, V. S., 7–8
Krueger, K., 115
Krutka, D. G., 97, 98–99

L
Laat, M., 99
Ladson-Billings, Gloria, 91
Laferrière, T., 2
LaPierre, J., 101–102
Latino parent tech nights, 13–14
learners. *See* students
learning
    adaptive technologies, 53
    flipped, 8, 51–52
    game-based, 55–56
    lifelong, 90
    online, 50–55
    virtual professional, 99
learning analytics, 53, 99–102
learning management systems (LMS), 50–51
Lego Mindstorms, 76
Lego WeDo, 59
library hours, extended, 12–13
lifelong learning, 90
likemindedness, 98
Lim, M., 43, 49
Liu, S., 43, 49
livestreaming, 103–107
Local Educational Agency, 109

M
MacArthur Foundation, 96
Magiera, Jennie, 9
makerspaces, 58–63
Makey Makeys, 59, 61–62, 63
massive open online courses (MOOCs), 54–55
mentors, virtual, 83
Merge Cube VR, 58
microcredentialing, 96–97
mixed reality, 57–58
mobile devices
    apps, 37
    "bring your own device" programs, 36
    Horizon Report, 35–36, 37
    smartphones, 35, 37
    tablets, 35–36, 37
mobile phones, 35, 37
modeling best practices, 10–11
motivational innovations
    about, 55
    augmented reality and virtual reality, 56–58
    educational gaming, 55–56
    makerspaces, 58–63

N
National Center for Education Statistics, 19–20
National Education Technology Plan, 15–16, 32, 56
National Telecommunications and Information Administration, 19
needs of teachers, 112–113
New Media Consortium, 49, 54. *See also* Horizon Report
New York's University at Buffalo, 58

Index

North Carolina Department of
Education, 107–108

O
Oakland Unified School District, 108
#ObserveMe, 106
online learning
about, 50
adaptive learning and learning
analytics, 53
flipped learning, 51–52
Horizon Report, 50, 51, 53, 54
learning management systems,
50–51
massive open online courses,
54–55
open educational resources, 54
open educational resources, 54
Open Letters
digital equity, 26–27
digital natives *versus* digital
immigrants, 93–95
diversity, 70–71
opportunity gaps overview, 9–10

P
parental permission for video/
livestreaming, 106–107
Paris, D., 91–92
partnerships, 107–109, 112
#PasstheScopeEDU, 105–106
Periscope, 105, 107
personal broadcasting, 43–44
personalization, 90
privacy, student, 106–107
professional learning networks
(PLNs), 90, 97–99
Project Tomorrow, 11, 25, 90
publicity releases, 106

Q
quality teaching, 16–18, 75–80
Quan-Lorey, Stephanie, 60–63

R
race, and device access, 20
racial bias, 73
Resta, P., 2
Riccards, P., 90
Rideout, V., 7–8
robotics, 59, 76
rural areas, 21–22
Ryerse, M., 96

S
San Jose State University, 110–111
Santa Clara County Office of
Education, 110–111
Sauceda Curwen, M., 72
Schaffer, Regina, 66
Schreurs, B., 99
Scratch (programming language),
60–62
Sheldon, J. P., 72
smartphones, 35, 37
social media, 38–43, 97–99
Solomon, G., 5
South, Joseph, 91
Speak Up Survey, 11, 25
Spirrison, Brad, 97, 98
Sprint 1Million Project grant, 13
State Educational Agency (SEA),
108–109
students
cultural background, 28–29,
91–92
ISTE Standards for Students,
78–79
privacy, 106–107
summer access reading project, 14
surveys, 11, 14, 25, 74–75

T
tablets, 35–36, 37
Teach Tomorrow in Oakland, 108
teacher preparation, critical issues in
about, 15–16
access to internet/devices, 18–24

Index

implementation plans, 24–25
quality teaching, 16–18
teacher preparation, non-traditional
    about, 95–96
    badges and microcredentialing,
        96–97
    Edcamps and unconferences, 99,
        102–103
    learning analytics, 99–102
    partnerships, 107–109, 112
    professional learning networks,
        97–99
    video and livestreaming, 103–107
teacher preparation, traditional
    advocate role, 91–92
    information, critical consumers of,
        90–91
    lifelong learning, 90
    personalization, 90
teachers
    ISTE Certification for Educators,
        109, 112
    ISTE Standards for Educators,
        3–4, 73, 77, 78–79
    needs, 112–113
    tech knowledge, assumptions
        about, 74–75
teaching, quality, 16–18, 75–80
technology
    new uses *versus* new tools,
        64–65
    as tool, 66–67
    wearable, 64
technology adoption predictions,
    33–35, 63–65. *See also*
    Horizon Report
Thomas, Sarah, 28, 52, 90, 98, 102,
    104–105, 106–107
Trayburn, D., 72
Twitter, 39, 41, 42, 65, 97–99. *See
    also* social media

U
unconferences, 102–103
University of Redlands, 58
U.S. Department of Education, 16,
    19, 89, 108–109
U.S. Department of Education Office
    for Civil Rights, 10
U.S. Department of Education Office
    of Educational Technology,
    1–2, 11

V
video, 43–44, 64, 103–107
virtual mentors, 83
virtual professional learning, 99
virtual reality, 56–58
Voithofer, R., 23

W
Wagner, K., 99
Watson, Jill (AI creation), 99–100
wearable technology, 64
Wi-Fi access maps, 13
William and Ida Friday Institute,
    107–108

Closing the Gap   **137**

Your Opinion Matters
Tell Us How We're Doing!

Your feedback helps ISTE create the best possible resources for teaching and learning in the digital age. Share your thoughts with the community or tell us how we're doing!

You Can:

- Write a review at amazon.com or barnesandnoble.com.
- Mention this book on social media and follow ISTE on Twitter @iste, Facebook @ISTEconnects or Instagram @isteconnects
- Email us at books@iste.org with your questions or comments.